PHOTONIC CRYSTALS - INNOVATIVE SYSTEMS, LASERS AND WAVEGUIDES

Editor

Eden Slaver

AF214464

scitus academics

Photonic Crystals - Innovative Systems, Lasers and Waveguides
Edited by **Eden Slaver**

ISBN: 978-1-68117-600-0
Library of Congress Control Number: 2016946455

© 2017 by
SCITUS Academics LLC,
616, Corporate Way, Suite 2, 4766,
Valley Cottage, NY 10989

www.scitusacademics.com

This book contains information obtained from highly regarded resources. Copyright for individual articles remains with the authors as indicated. All chapters are distributed under the terms of the Creative Commons Attribution License, which permits unrestricted use, distribution, and reproduction in any medium, provided the original author and source are credited.

Notice

Reasonable efforts have been made to publish reliable data and views articulated in the chapters are those of the individual contributors, and not necessarily those of the editors or publishers. Editors or publishers are not responsible for the accuracy of the information in the published chapters or consequences of their use. The publisher believes no responsibility for any damage or grievance to the persons or property arising out of the use of any materials, instructions, methods or thoughts in the book. The editors and the publisher have attempted to trace the copyright holders of all material reproduced in this publication and apologize to copyright holders if permission has not been obtained. If any copyright holder has not been acknowledged, please write to us so we may rectify.

Printed in United States of America on Acid Free Paper

PREFACE

A photonic crystal is a regular optical nanostructure that affects the movement of photons in the same way that ionic lattices affect electrons in solids very exactly. Photonic crystals occur in nature in the form of structural coloring-and promise to be useful in a variety of applications in various forms. In principle, photonic crystals can be used wherever light needs to be handled. Current uses include thin film optics with lens coatings. In non-linear devices, two-dimensional photonic crystal fibers are used to guide exotic wavelengths. In optical computers, three-dimensional crystals can be used one day. This book includes the characterization of photonic crystals, photonic crystal lasers, photonic crystal waveguides and plasmonics, including innovative systems and materials. Photonic crystal materials could also be utilized for the production of ultra-low power sources of light. Researchers have studied lasers in photonic crystals from microscopic cavities that act as reflectors to intensify the collisions between photons and atoms that lead to lazing, but these lasers have been optically pumped, which means they are driven by other lasers.

TABLE OF CONTENTS

1

HARMONIC MODE-LOCKED FIBER LASER BASED ON PHOTONIC CRYSTAL FIBER FILLED WITH TOPOLOGICAL INSULATOR SOLUTION

Yu-Shan Chen, Pei-Guang Yan, Hao Chen, Ai-Jiang Liu and Shuang-Chen Ruan *

Shenzhen Key Laboratory of Laser Engineering, Key Laboratory of Advanced Optical Precision Manufacturing Technology of Guangdong Higher Education Institutes, College of Optoelectronic Engineering, Shenzhen University, Shenzhen, Guangdong 518060, China

ABSTRACT

We reported that the photonic crystal fiber (PCF) filled with TI:Bi_2Te_3 nanosheets solution could act as an effective saturable absorber (SA). Employing this TI-PCF SA device; we constructed an ytterbium-doped all-fiber laser oscillator and achieved the evanescent wave mode-locking operation. Due to the large cavity dispersion; the fundamental mode-locking pulse had the large full width at half maximum (FWHM) of 2.33 ns with the repetition rate of ~1.11 MHz; and the radio frequency (RF) spectrum with signal-to-noise ratio (SNR) of 61 dB. In addition; the transition dynamics from a bunched state of pulses to harmonic mode-locking (HML) was also observed; which was up to 26th order.

KEYWORDS

Topological Insulator; photonic crystal fiber; mode-locked fiber laser; evanescent wave; harmonic mode-locking

1. INTRODUCTION

In the past decade, passively mode-locked fiber lasers have been extensively investigated due to its simplicity and compactness [1,2,3]. The high repetition rates of passively mode-locked fiber lasers with subnanosecond to picosecond pulses have potential applications in many fields, including high speed optical communication, optical frequency metrology [4,5]. As an important technique to increase the pulse repetition rate, the relevant harmonic mode locking (HML) fiber lasers have been a growing research region. To date, most of the previous works on HML were related to the nonlinear polarization rotation (NPR) [6,7,8,9] and nonlinear amplifying mirrors (NALMs) technique [10]. However, mode-locked fiber lasers based on NPR and the NALM technique are intrinsically environmentally unstable. Therefore, researchers have strong motivations investigate HML in passively mode-locked fiber lasers by real saturable absorbers (SAs) [11,12,13,14,15,16], such as semiconductor saturable absorbers (SESAM), single-walled carbon nanotube (SWNT) and graphene. However, these SAs also possess some inherent drawbacks. SESAMs have a narrow broadband response and require complex fabrication process and expensive equipment. SWNTs are also restricted by their diameter-dependent response spectral range, and suffered with a relatively high non-saturable loss for broadband operation. In addition, single-layer graphene is an intrinsic ultra-broadband SA, but its modulation depth is often low (typically <1% per layer [17]). Also, grapheme powder [18] and nanoscale charcoal powder [19] were employed as nano-scale saturable absorbers effectively achieve mode-locked.

Recently, another new type of Dirac material, topological insulators (TIs), which have become one of the most interesting research topics in condensed-matter physics [20,21]. At the same time, they have been successfully utilized as SA for the realization of mode-locking [22,23,24,25,26,27,28,29,30,31,32,33,34,35,36,37,38,39,40,41]. TISA could be fabricated by the mechanical exfoliation, the molecular beam epitaxy (MBE) growth, the vapor-liquid-solid growth, chemical exfoliation, the mechanical triturating and the filmy method in polyvinyl alcohol [25,42,43,44]. In most of recent studies, the nanostructured TIs are typically deposited onto the fiber ferrule [23,28,29,45]. However, the interaction length of the nanostructured TI with light is limited due to large material losses. The SA device based on the evanescent wave interaction is another effective approach to achieve mode-locked fiber laser [22,32,46,47]. Nevertheless, the microfiber-based SAs are very frangible with the disadvantage of high insertion loss (IL) and short interaction length [27], the side-polished-fiber-based SAs exhibit polarization dependent loss due to centrosymmetric structure [48]. Photonic crystal fiber (PCF) [49], characterized with regular air holes along fiber length, has been utilized as platform of SA for evanescent wave interaction by filling the nanoparticles into cladding holes [50,51,52]. More recently, the HML erbium-doped fiber lasers based on TISA has also been investigated experimentally [27,53]. However, no previous experiments have reported the HML in an ytterbium-doped all-fiber laser oscillator in the all-normal dispersion regime based on TI-filled PCF (TI-PCF) SA. In addition, it has been demonstrated that introducing proper high nonlinear effect into the laser cavity was favorable for generating harmonic mode locking (HML) pulses [27]. The small-core PCF usually has a higher nonlinearity than conventional single mode fiber (SMF). Meanwhile, TIs possess giant third order nonlinear optical property (~10–14 m^2/W) [54]. Therefore, the TI-PCF type of saturable absorber can introduce larger nonlinearity, benefiting for the generation of HML pulses.

In this paper, we reported that the PCF filled with TI:Bi_2Te_3 nanosheets solution could act as an effective SA. Employing this TI-

PCF SA device, we constructed an ytterbium-doped all-fiber laser oscillator and achieved the evanescent wave mode-locking operation. Fundamental mode-locking and harmonic mode-locking (HML) up to 26th order were demonstrated.

2. SAMPLE PREPARATION AND EXPERIMENTAL SETUP

The scanning electron micrographs (SEM) of the PCF were shown in Figure 1. The total outer diameter was measured about 127 μm. The central core, fabricated by replacing the central air hole with a silica rod, had a diameter of 9.5 μm. The diameter and pitch of air holes were 5.62 μm and 8.3 μm, respectively. The fundamental and 2nd mode distribution were calculated and shown in the insets of Figure 1b. The guided light in fundamental mode could interact with the TI:Bi_2Te_3nanosheets in solution through the evanescent wave. Acting as the platform of SA, it should be pointed out that the PCF could eliminate the higher order mode in laser cavity under the following conditions: (1) the PCF was deliberately bent in our experiment tightly to suppress the high order modes; (2) Both ends of the PCF was spliced with SMF-28 fiber to avoid the excitation of high order mode. The mode field diameter (MFD) of the PCF was measured ~8.7 μm at 1310 nm. And its attenuation and dispersion parameter at 1060 nm were measured to be 2.2 dB/km and –20.5 ps/km/nm, respectively.

The TI:Bi_2Te_3 nanosheets were synthesized with a low cost and convenient approach based on the hydrothermal intercalation/exfoliation method. Figure 2a showed the SEM of TI:Bi_2Te_3 nanosheets. The TI solution was shown in the inset of Figure 2b, which fabricated by dispersing 0.25 mg samples into 4 mL deionized water, and homogenized with 30-minites ultrasonication. The transmission spectrum of the solution was measured from 1000 nm to 1700 nm using an optical spectrometer (Perkinelmer Lambda

7500), and the linear-transmission ratio was ~43% at 1060 nm, as shown in Figure 2b.

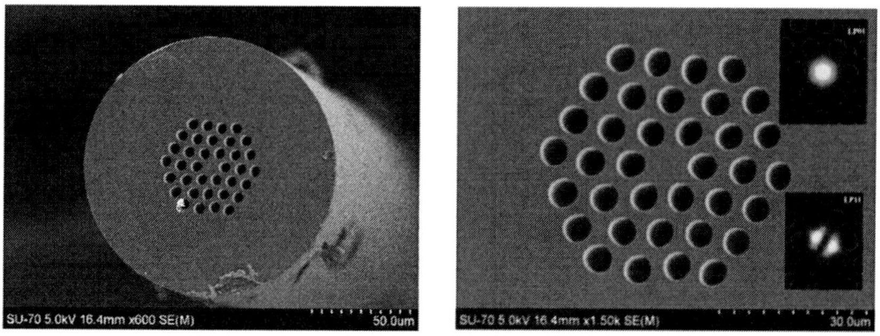

(a) (b)

Figure 1. The structure of PCF. (**a**) the whole cross section, (**b**) the enlarged microstructured region. The insets in (**b**) were calculated mode distribution of fundamental mode and 2nd mode.

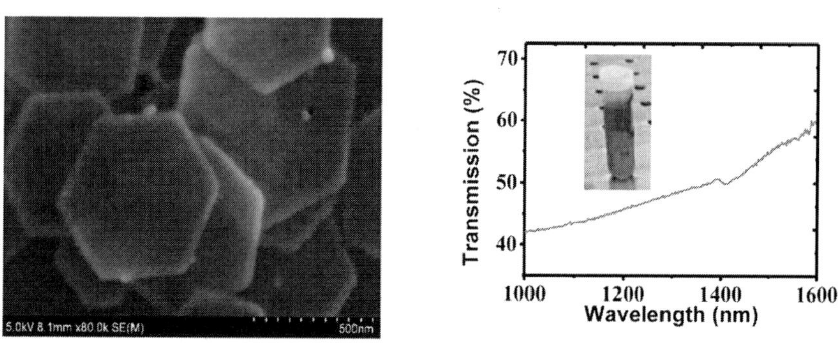

(a) (b)

Figure 2. (**a**) SEM of TI:Bi_2Te_3 nanosheets; (**b**) The transmission spectrum of the solution.

The TI solution was pulled into the air holes of PCF by an injector. The TI-PCF SA device had a length of 8cm with each end spliced with SMF-28 fiber. The measured insertion loss (IL) was only ~0.45 dB. To check the TI nanosheets really modulated the guided light in PCF, the saturable absorption property of the TI-PCF SA was measured and shown in Figure 3.

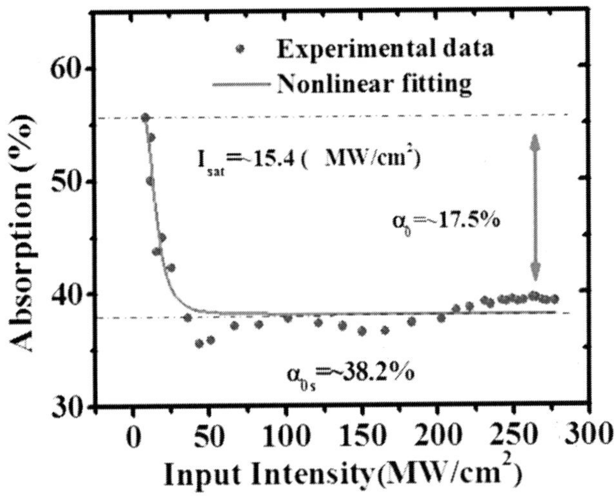

Figure 3. Measured nonlinear absorbance of the TI-PCF SA device.

The laser source used was homebuilt mode-locked oscillator with 5 MHz repetition rate and 150 ps pulse duration, which operating at ~1064 nm. The date obtained from the experiment was then fitted according to

$$\alpha(I) = \alpha_0 \left(1 + \frac{I}{I_{sat}}\right)^{-1} + \alpha_{ns} \tag{1}$$

where I was the input laser intensity, I_{sat} was the saturable intensity, $\alpha(I)$ was the intensity-dependent absorption coefficient, and α_0 and

α_{ns} were the modulation depth and the non-saturable loss, respectively. The results gave a saturable intensity of ~15.4 MW/cm^2, a modulation depth of ~17.5%, and a non-saturable loss of ~38.2%. The saturable intensity of the TI-PCF SA was comparative with 53 MW/cm^2 of TI: Bi_2Se_3 deposited onto fiber end facet [24] and 12 MW/cm^2 of TI: Bi_2Se_3 embedded into PVA film [30]. On the other hand, the modulation depth of the TI-PCF SA would be high enough for stable femtosecond pulse mode-locking, considering that another TI SA with 1.7% modulation depth has demonstrated the HML according to a recent report by Z. Luo *et al.* [27].

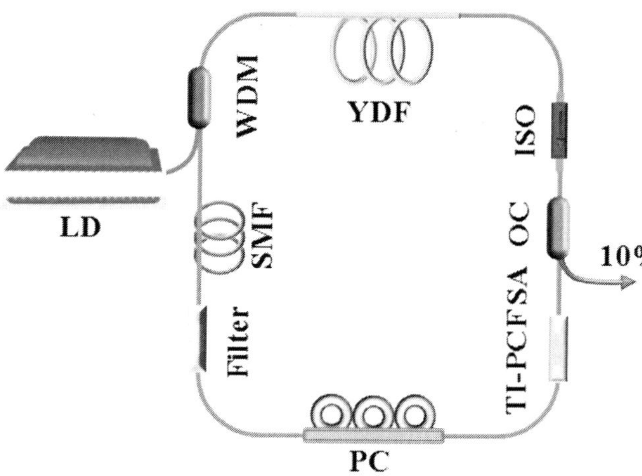

Figure 4. Schematic diagram of the fiber ring laser.

Our fiber laser was schematically shown in Figure 4. The laser was pumped by a 976 nm pump laser through a wavelength division multiplexer (WDM), and the output of the laser was coupled through a 10% fiber optical coupler (OC). A segment of ~3.4 m YDF with absorption of 250 dB/m@975 nm was used as the gain medium. A polarization independent isolator (ISO) was used to ensure the unidirectional operation of the ring laser cavity, and a three-spool polarization controller (PC) was employed to adjust the cavity birefringence. A bandwidth filter with a central wavelength of 1064 nm and 3 dB bandwidth of ~5 nm was inserted into the cavity. A total length of ~180 m HI-1060 Flex single mode fiber (SMF) was also incorporated to the cavity. The total cavity length was ~185 m. The monitoring of the output temporal pulse trains and optical spectrum were performed using a 20-GHz mixed signal oscilloscope (Tektronix MSO72004C) with a 45-GHz photo-detector (Newport 1014) and an optical spectrum analyzer (OSA, AQ6370B) with a minimum resolution of 0.02 nm, respectively. The pump power and output power were monitored by a photodiode power meter (COHERENT). In addition, the radio frequency (RF) spectrum of the fundamental mode-locking was measured by a Mixed Domain Oscilloscope (Tektronix MDO4034B-3).

3. EXPERIMENTAL RESULTS AND DISCUSSION

3.1. Fundamental Mode-Locking

Continuous wave operation occurred at a pump power of ~76 mW. The stable operation in fundamental mode locking was observed up to the ~112 mW of pump power after careful adjustment of the PC, as shown in Figure 5a. The pulse train had a fundamental cavity frequency ~1.11 MHz, which was determined by the ~185 m cavity length. As depicted in the inset ofFigure 5a, the full width at half maximum (FWHM) of the single pulse was ~2.33 ns. Figure 5b showed the typical spectrum of mode-locked pulses, it was centered at ~1062.27 nm with a 3 dB bandwidth of ~0.80 nm. In addition, the typical characteristics of steep spectral edges indicated that the mode-locked pulses have been shaped to dissipative solitons, which are found exclusively in all normal-dispersion lasers, such as ytterbium-doped all-fiber laser [55]. The corresponding RF spectrum of the laser was shown Figure 5c, and it was clearly shown that the repetition rate of the pulse train was ~1.11 MHz. The electrical signal-to-noise ratio (SNR) was ~61 dB measured with 20 kHz resolution bandwidth (RBW). It was introduced more additional linear dispersion and nonlinearity into the all-normal-dispersion cavity due to the extra fiber of ~185 m, which required for the extreme pulse broadening [56,57]. Thus, it was reasonable that the single pulse here was larger than 2 ns.

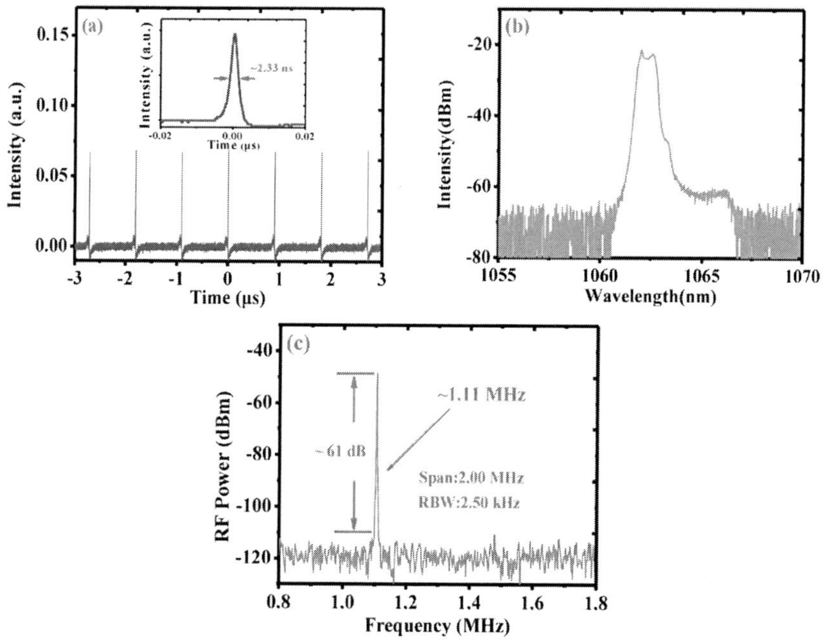

Figure 5. (**a**) Fundamental mode-locking pulse train. Inset: single pulse width. (**b**) Corresponding spectrum. (**c**) Fundamental frequency signal.

3.2. Harmonic Mode-Locking

For a passively harmonic mode-locked laser, splitting of pulses usually occurred before the attainment of nearly equal-spaced and self-organized pulse train. By increasing the pump power and adjusting the orientation of PC carefully, splitting of pulse occurred owing to the cavity pulse peak clamping effect. This process was similar to the soliton operation state in the fiber lasers with net negative cavity group-velocity dispersion [58]. The multiple pulses then rearranged themselves automatically, and all pulses were equally spaced in the cavity. Firstly, 2nd-order HML was observed at the pump power of ~127 mW by changing the PC, as shown in Figure 6a. The corresponding RF spectrum was measured, which was given inFigure 6e. It could be seen that the SNR of 2nd order HML was ~ 45 dB, and its repetition rate was ~2.21 MHz (RBW was 2.50 kHz). The harmonic number could also be enlarged only by increasing the pump power slightly in our experiment, the pulse trains of the 3rd (Figure 6b), 4th (Figure 6c) and 5th (Figure 6d) order HML could be obtained under the pump powers of ~131 mW, ~140 mW, ~149 mW, respectively. In addition, the corresponding RF spectra of the 3rd (the repetition rate of ~3.32 MHz), 4th (the repetition rate of ~4.42 MHz) and 5th (the repetition rate of ~5.53 MHz) order HML were given inFigure 6f–h with the corresponding SNRs of ~ 43 dB, ~ 42 dB and ~ 42 dB, respectively. Finally, an inverse progress was explored. When the pump power was decreased to ~137 mW, the 4th order HML appeared again, then the 3th and 2nd order HML appeared in turn under a little below the previous pump powers due to the hysteresis phenomena of HML [59]. Similar to the previous work [60], the multi-wavelength mode-locked pulses were also observed in the experiment.

Further increasing the pump power to ~324 mW, a noticeable transition from a bunched state of pulses to high-order HML could be repeatedly achieved by tuning the orientation of PC, as illustrated in Figure 7a–e. It was clearly shown that the bunch of pulses expanded gradually and occupied the whole cavity regularly, leading to the formation of a high-order HML state eventually (shown in Figure 7e. The Figure 7f indicated that the FWHM of a single pulse in HML was ~575.80 ps.Figure 7g showed the typical output optical spectra of initial state (corresponding to the oscilloscope trace of Figure 7a and final state (corresponding to the oscilloscope trace of Figure 7e. The HML had a central wavelength of 1065.4 nm and a 3-dB spectral bandwidth of 1.51 nm. The corresponding RF spectrum was also measured, as shown in Figure 7h. The SNR of the high-order HML was ~ 38 dB, and the final repetition rate was ~28.73 MHz, corresponding to the 26th harmonic. Here we use photonic crystal fiber filled with TI:Bi_2Te_3 nanosheets solution as saturable absorber to achieve harmonic mode-locking. Another way [61] like single-wall carbon nanotube (SWCNT)-doped polyvinyl alcohol (PVA)-based as saturable absorber through using either all-physical-contact (PC) or all-angled-physical-contact (APC) connectors can also be used.

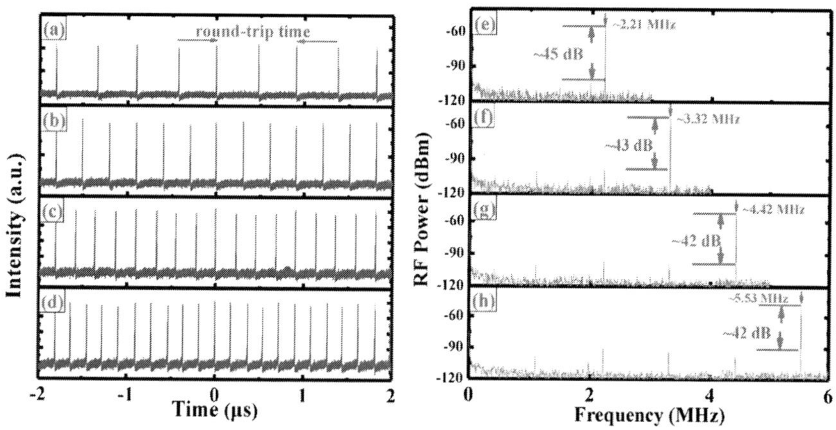

Figure 6. Pulse trains for (**a**) 2nd HML, (**b**) 3rd HML, (**c**) 4th HML and (**d**) 5th HML as well as the corresponding RF-spectra for (**e**) 2nd HML, (**f**) 3rd HML, (**g**) 4th HML and (**h**) 5th HML, respectively.

4. CONCLUSIONS

As a summary, a TI-PCF SA device was firstly demonstrated, which possessed some unique advantages, such as single mode operation, nonlinearity enhancement, low insertion loss, all-fiber configuration, longer action length. Employing this SA device, an ytterbium-doped all-fiber laser was constructed and the evanescent wave mode-locking operation was achieved. Not only stable fundamental mode-locking and low order HML, but also the transition dynamics from a bunched state of pulses to high order HML were demonstrated. Our experimental results demonstrate a new scheme for ultra-fast laser mode-locking application.

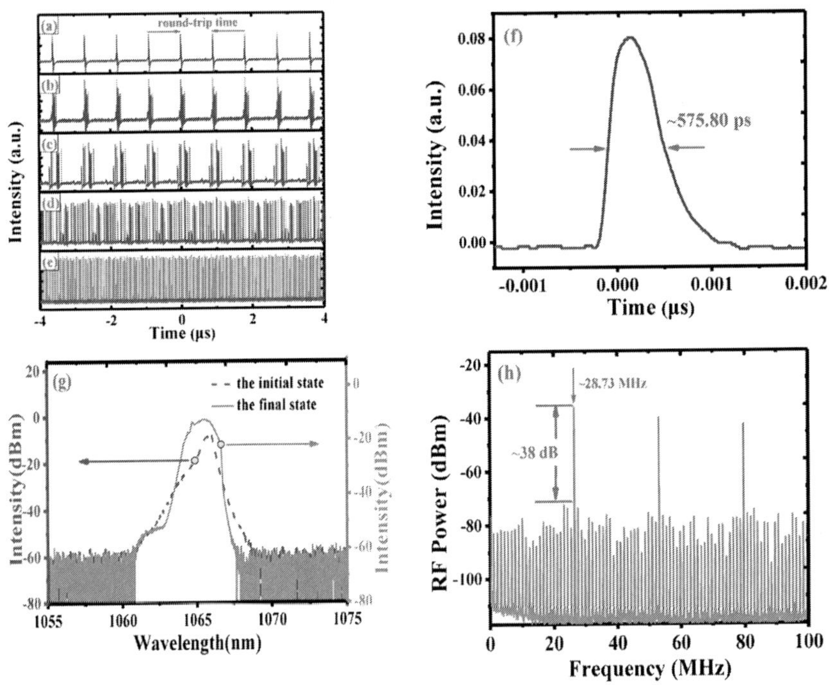

Figure 7. (**a–e**) the transition from a bunched state to the 26th-order HML. (**f**) the single pulse of 26th-order HML. (**g**) Optical spectra of initial and final state. (**h**) The corresponding RF spectrum of 26th-order HML.

ACKNOWLEDGMENTS

This research was Supported by the NSFC (61275144), Natural science fund of Guangdong province (S2013010012235), the foundation for scientific and technical innovation in Higher Education of Guangdong (2013KJCX0161), the Improvement and Development Project of Shenzhen Key Lab (ZDSY20120 612094924467), the Science and technology project of Shenzhen City (JCYJ20120613172042264, JCYJ201303291420407 31), Natural Science Foundation of SZU (No.201221).

AUTHOR CONTRIBUTIONS

Writing of paper: Yu-Shan Chen. Comments and feedback: Yu-Shan Chen. Experimental investigations: Pei-Guang Yan, Yu-Shan Chen, Hao Chen, Ai-Jiang Liu and Shuang-Chen Ruan. Schemes and design of investigations: Pei-Guang Yan.

CONFLICT OF INTEREST

The authors declare no conflict of interest.

REFERENCES

1. Zhang, H.; Tang, D.; Knize, R.J.; Zhao, L.; Bao, Q.; Loh, K.P. Graphene mode locked, wavelength-tunable, dissipative soliton fiber laser. Appl. Phys. Lett. 2010, 96, 111–112.

2. Wise, F.W.; Chong, A.; Renninger, W.H. High-energy femtosecond fiber lasers based on pulse propagation at normal dispersion. Laser Photon. Rev. 2008, 2, 58–73.

3. Kobtsev, S.; Kukarin, S.; Fedotov, Y. Ultra-low repetition rate mode-locked fiber laser with high-energy pulses. Opt. Express 2008, 16, 21936–21941.

4. McFerran, J.J. Echelle spectrograph calibration with a frequency comb based on a harmonically mode-locked fiber laser: A proposal. Appl. Opt. 2009, 48, 2752–2759.

5. Mikulla, B.; Leng, L.; Sears, S.; Collings, B.C.; Arend, M.; Bergman, K. Broad-Band High-Repetition-Rate Source for Spectrally Sliced WDM. IEEE Photon. Technol. Lett. 1999, 11, 418–420.

6. Zhang, Z.X.; Zhan, L.; Yang, X.X.; Luo, S.Y.; Xia, Y.X. Passive harmonically mode-locked erbium-doped fiber laser with scalable repetition rate up to 1.2 GHz. Laser Phys. Lett. 2007, 4, 592.

7. Amrani, F.; Haboucha, A.; Salhi, M.; Leblond, H.; Komarov, A.; Grelu, Ph.; Sanchez, F. Passively mode-locked erbium-doped double-clad fiber laser operating at the 322nd harmonic. Opt. Lett. 2009, 34, 2120–2122.

8. Zhu, X.J.; Wang, C.H.; Liu, S.X.; Zhang, G.J. Tunable High-Order Harmonic Mode-Locking in Yb-Doped Fiber Laser With All-Normal Dispersion. Photon. Technol. Lett. 2012, 24, 754–756.

9. Peng, J.; Zhan, L.; Luo, S.; Shen, Q. Passive Harmonic Mode-Locking of Dissipative Solitons in a Normal-Dispersion Er-Doped Fiber Laser. J. Lightw. Technol. 2013, 31, 3009–3014.

10. Chen, H.R.; Lin, K.H.; Tsai, C.Y.; Wu, H.H.; Wu, C.H.; Chen, C.H.; Chi, Y.C.; Lin, G.R.; Hsieh, W.F. 12 GHz passive harmonic mode-locking in a 1.06 μm semiconductor optical amplifier-based fiber laser

with figure-eight cavity configuration. Opt. Lett. 2013, 38, 845–847.

11. Jun, C.S.; Choi, S.Y.; Rotermund, F.; Kim, B.Y.; Yeom, D. Toward higher-order passive harmonic mode-locking of a soliton fiber laser. Opt. Lett. 2012, 37, 1862–1864.

12. Jun, C.S.; Im, J.H.; Yoo, S.H.; Choi, S.Y.; Rotermund, F.; Yeom, D.; Kim, B.Y. Low noise GHz passive harmonic mode-locking of soliton fiber laser using evanescent wave interaction with carbon nanotubes. Opt. Express 2011, 19, 19775–19780.

13. Mou, C.B.; Arif, R.; Rozhin, A.; Turitsyn, S. Passively harmonic mode locked erbium doped fiber soliton laser with carbon nanotubes based saturable absorber. Opt. Mater. Express 2012, 2, 884–890.

14. Sotor, J.; Sobon, G.; Macherzynski, W.; Abramski, K.M. Harmonically mode-locked Er-doped fiber laser based on a Sb2Te3 topological insulator saturable absorber. Laser Phys. Lett. 2014, 11, 055102.

15. Sobon, G.; Sotor, J.; Abramski, K.M. Passive harmonic mode-locking in Er-doped fiber laser based on graphene saturable absorber with repetition rates scalable to 2.22 GHz. Appl. Phys. Lett. 2012, 100, 161109.

16. Huang, S.S.; Wang, Y.G.; Yan, P.G.; Zhang, G.L.; Zhao, J.Q.; Li, H.Q.; Lin, R.Y. High order harmonic modelocking in an allnormaldispersion Ybdoped fiber laser with a graphene oxide saturable absorber. Laser Phys. 2014, 24, 015001.

17. Martinez, A.; Sun, Z. Nanotube and graphene saturable absorbers for fibre lasers. Nat. Photon. 2013, 7, 842–845.

18. Lin, G.; Lin, Y. Directly exfoliated and imprinted graphite nano-particle saturable absorber for passive mode-locking erbium-doped fiber laser. Laser Phys. Lett. 2011, 8, 880.

19. Lin, Y.H.; Chi, Y.C.; Lin, G.R. Nanoscale charcoal powder induced saturable absorption and mode-locking of a low-gain erbium-doped fiber-ring laser. Laser Phys. Lett. 2013, 10, 055105.

20. Qi, X.L.; Zhang, S.C. Topological insulators and superconductors. Rev. Mod. Phys. 2011, 83, 1057.

21. Hasan, M.Z.; Kane, C.L. Colloquium: Topological insulators. Rev. Mod. Phys. 2010, 82, 3045.

22. Sotor, J.; Sobon, G.; Abramski, K.M. Sub-130 fs mode-locked er-doped fiber laser based on topological insulator. Opt. Express 2014, 22, 13244–13249.

23. Zhao, C.; Zhang, H.; Qi, X.; Chen, Y.; Wang, Z.; Wen, S.; Tang, D. Ultra-short pulse generation by a topological insulator based saturable absorber. Appl. Phys. Lett. 2012, 101, 211106.

24. Luo, Z.; Huang, Y.; Weng, J.; Cheng, H.; Lin, Z.; Xu, B.; Cai, Z.; Xu, H. 1.06 μm q-switched ytterbium-doped fiber laser using few-layer topological insulator Bi2Se3 as a saturable absorber. Opt. Express 2013, 21, 29516–29522.

25. Yu, Z.; Song, Y.; Tian, J.; Dou, Z.; Guoyu, H.; Li, K.; Li, H.; Zhang, X. High-repetition-rate q-switched fiber laser with high quality topological insulator Bi2Se3 film. Opt. Express 2014, 22, 11508–11515.

26. Chen, Y.; Zhao, C.; Huang, H.; Chen, S.; Tang, P.; Wang, Z.; Lu, S.; Zhang, H.; Wen, S.; Tang, D. Self-Assembled Topological Insulator: Bi2Se3 Membrane as a Passive Q-Switcher in an Erbium-Doped Fiber Laser. J. Lightw. Technol.2013, 31, 2857–2863.

27. Luo, Z.C.; Liu, M.; Liu, H.; Zheng, X.W.; Luo, A.P.; Zhao, C.J.; Zhang, H.; Wen, S.C.; Xu, W.C. 2 GHZ passively harmonic mode-locked fiber laser by a microfiber-based topological insulator saturable absorber. Opt. Lett. 2013, 38, 5212–5215.

28. Zhao, C.; Zou, Y.; Chen, Y.; Wang, Z.; Lu, S.; Zhang, H.; Wen, S.; Tang, D. Wavelength-tunable picosecond soliton fiber laser with topological insulator: Bi2Se3 as a mode locker. Opt. Express 2012, 20, 27888–27895.

29. Luo, Z.; Liu, C.; Huang, Y.; Wu, D.; Wu, J.; Xu, H.; Cai, Z.; Lin, Z.; Sun, L.; Weng, J. Topological-insulator passively q-switched double-clad fiber laser at 2 µm wavelength. IEEE J. Sel. Top. Quant. Electron. 2014, 20, 0902708.

30. Liu, H.; Zheng, X.W.; Liu, M.; Zhao, N.; Luo, A.P.; Luo, Z.C.; Xu, W.C.; Zhang, H.; Zhao, C.J.; Wen, S.C. Femtosecond pulse generation from a topological insulator mode-locked fiber laser. Opt. Express 2014, 22, 6868–6873.

31. Chen, Y.; Zhao, C.; Chen, S.; Du, J.; Tang, P.; Jiang, G.; Zhang, H.; Wen, S.; Tang, D. Large energy, wavelength widely tunable, topological insulator q-switched erbium-doped fiber laser. IEEE J. Sel. Top. Quant. Electron. 2014, 20, 315–322.

32. Lee, J.; Koo, J.; Jhon, Y.-M.; Lee, J.H. A femtosecond pulse erbium fiber laser incorporating a saturable absorber based on bulk-structured Bi2Se3 topological insulator. Opt. Express 2014, 22, 6165–6173.

33. Lin, Y.H.; Yang, C.Y.; Lin, S.F.; Tseng, W.H.; Bao, Q.; Wu, C.-I.; Lin, G.-R. Soliton compression of the erbium-doped fiber laser weakly started mode-locking by nanoscale p-type Bi2Te3 topological insulator particles. Laser Phys. Lett.2014, 11, 055107.

34. Yan, P.; Lin, R.; Ruan, S.; Liu, A.; Chen, H.; Zheng, Y.; Chen, S.; Guo, C.; Hu, J. A practical topological insulator saturable absorber for mode-locked fiber laser. Sci. Rep. 2015, 5, 8690.

35. Yan, P.; Lin, R.; Ruan, S.; Liu, A.; Chen, H. A 2.95 GHz, femtosecond passive harmonic mode-locked fiber laser based on evanescent

field interaction with topological insulator film. Opt. Express. 2015, 23, 154–164.

36. Yan, P.; Lin, R.; Chen, H.; Zhang, H.; Liu, A.; Yang, H.; Ruan, S. Topological insulator solution filled in photonic crystal fiber for passive mode-locked fiber laser. IEEE Photon. Tech. Lett. 2015, 27, 264–267.

37. Huang, S.; Wang, Y.; Yan, P.; Zhao, J.; Li, H.; Lin, R. Tunable and switchable multi-wavelength dissipative soliton generation in a graphene oxide mode-locked Yb-doped fiber laser. Opt. Express 2014, 22, 11417–11426.

38. Lin, R.Y.; Wang, Y.G.; Yan, P.G.; Zhang, G.-L.; Zhao, J.-Q.; Li, H.Q.; Huang, S.S.; Cao, G.Z.; Duan, J.-A. Bright and dark square pulses generated from a graphene-oxide mode-locked ytterbium-doped fiber laser. Photon. J. IEEE. 2014, 6, 1–8.

39. Huang, S.; Wang, Y.; Yan, P.; Zhao, J.; Li, H.; Lin, R. Tunable and switchable multi-wavelength dissipative soliton generation in a graphene oxide mode-locked Yb-doped fiber laser. Optics express 2014, 22, 11417–11426.

40. Yan, P.; Liu, A.; Chen, Y.; Chen, H.; Ruan, S.; Guo, C.; Chen, S.; Li, I.; Yang, H.; Hu, J.; Cao, G. Microfiber-based WS2-film saturable absorber for ultra-fast photonics. Opt. Mater. Express 2015, 5, 479–489.

41. Boguslawski, J.; Sotor, J.; Sobon, G.; Tarka, J.; Jagiello, J.; Macherzynski, W.; Lipinska, L.; Abramski, K.M. Mode-locked er-doped fiber laser based on liquid phase exfoliated Sb2Te3 topological insulator. Laser Phys. 2014, 24, 105111.

42. Bernard, F.; Zhang, H.; Gorza, S.P.; Emplit, P. Towards mode-locked fiber laser using topological insulators. In Proceedings of the Nonlinear Photonics, Colorado Springs, CO, USA, 17–21 June 2012.

43. Bansal, N.; Kim, Y.S.; Edrey, E.; Brahlek, M.; Horibe, Y.; Iida, K.; Tanimura, M.; Li, G.-H.; Feng, T.; Lee, H.-D. Epitaxial growth of topological insulator Bi2Se3 film on Si (111) with atomically sharp interface. Thin Solid Film. 2011, 520, 224–229.

44. Jung, M.; Lee, J.; Koo, J.; Park, J.; Song, Y.-W.; Lee, K.; Lee, S.; Lee, J.H. A femtosecond pulse fiber laser at 1935 nm using a bulk-structured Bi2Se3 topological insulator. Opt. Express 2014, 22, 7865–7874.

45. Chen, Y.; Wu, M.; Tang, P.; Chen, S.; Du, J.; Jiang, G.; Li, Y.; Zhao, C.; Zhang, H.; Wen, S. The formation of various multi-soliton patterns and noise-like pulse in a fiber laser passively mode-locked by a topological insulator based saturable absorber. Laser Phys. Lett. 2014, 11, 055101.

46. Zhao, N.; Liu, M.; Liu, H.; Zheng, X.-W.; Ning, Q.-Y.; Luo, A.-P.; Luo, Z.-C.; Xu, W.-C. Dual-wavelength rectangular pulse Yb-doped fiber laser using a microfiber-based graphene saturable absorber. Opt. Express 2014, 22, 10906–10913.

47. Luo, Z.; Zhou, M.; Wu, D.; Ye, C.; Weng, J.; Dong, J.; Xu, H.; Cai, Z.; Chen, L. Graphene-induced nonlinear four-wave-mixing and its application to multiwavelength q-switched rare-earth-doped fiber lasers. J. Lightw. Technol. 2011, 29, 2732–2739.

48. Choi, S.Y.; Cho, D.K.; Song, Y.-W.; Oh, K.; Kim, K.; Rotermund, F.; Yeom, D.-I. Graphene-filled hollow optical fiber saturable absorber for efficient soliton fiber laser mode-locking. Opt. Express 2012, 20, 5652–5657.

49. Knight, J.C.; Broeng, J.; Birks, T.A.; Russell, P.S.J. Photonic band gap guidance in optical fibers. Science 1998, 282, 1476–1478.

50. Liu, Z.B.; He, X.; Wang, D. Passively mode-locked fiber laser based on a hollow-core photonic crystal fiber filled with few-layered graphene oxide solution. Opt. Lett. 2011, 36, 3024–3026.

51. Lin, Y.H.; Yang, C.Y.; Liou, J.H.; Yu, C.P.; Lin, G.R. Using graphene nano-particle embedded in photonic crystal fiber for evanescent wave mode-locking of fiber laser. Opt. Express 2013, 21, 16763–16776.

52. Zhao, J.; Ruan, S.; Yan, P.; Zhang, H.; Yu, Y.; Wei, H.; Luo, J. Cladding-filled graphene in a photonic crystal fiber as a saturable absorber and its first application for ultrafast all-fiber laser. Opt. Eng. 2013, 52, 106105–106105.

53. Liu, M.; Zhao, N.; Liu, H.; Zheng, X.-W.; Luo, A.-P.; Luo, Z.-C.; Xu, W.-C.; Zhao, C.-J.; Zhang, H.; Wen, S.-C. Dual-wavelength harmonically mode-locked fiber laser with topological insulator saturable absorber. IEEE Photon. Technol. Lett. 2014, 26, 983–986.

54. Lu, S.B.; Zhao, C.J.; Zou, Y.H.; Chen, S.Q.; Chen, Y.; Li, Y.; Zhang, H.; Wen, S.C.; Tang, D.Y. Third order nonlinear optical property of Bi2Se3. Opt. Express 2013, 21, 2072–2082.

55. Grelu, P.; Akhmediev, N. Dissipative solitons for mode-locked lasers. Nat. Photon. 2012, 6, 84–92.

56. Kelleher, E.J.R.; Travers, J.C.; Ippen, E.P.; Sun, Z.; Ferrari, A.C.; Popov, S.V.; Taylor, J.R. Generation and direct measurement of giant chirp in a passively mode-locked laser. Opt. Lett. 2009, 34, 3526–3528.

57. Kelleher1, E.J.R.; Travers, J.C.; Sun, Z.; Rozhin, A.G.; Ferrari, A.C.; Popov, S.V.; Taylor, J.R. Nanosecond-pulse fiber lasers mode-locked with nanotubes. Appl. Phys. Lett. 2009, 95, 111108.

58. Tang, D.Y.; Zhao, L.M.; Zhao, B.; Liu, A.Q. Mechanism of multisoliton formation and soliton energy quantization in passively mode-locked fiber lasers. Phys. Rev. A. 2005, 72, 43816.

59. Liu, X. Hysteresis phenomena and multipulse formation of a dissipative system in a passively mode-locked fiber laser. Phys. Rev. A. 2010, 81, 23811.

60. Guo, B.; Yao, Y.; Tian, J.; Zhao, Y.; Liu, S.; Li, M.; Quan, M. Observation of bright-dark soliton pair in a mode-locked fiber laser with topological insulator. IEEE Photon. Technol. Lett. 2015, 27, 701–704.

61. Cheng, K.N.; Lin, Y.H.; Yamashita, S.; Lin, G.R. Harmonic order-dependent pulsewidth shortening of a passively mode-locked fiber laser with a carbon nanotube saturable absorber. Photonics Journal, IEEE. 2012, 4, 1542–1552.

2

DYNAMIC TUNING OF TRANSMISSION WAVELENGTH OF MEMS-BASED GE WAVEGUIDES ON A SI BEAM

Masashi Hirase, Luan M. Nguyen, Hiroshi Fukuda, Yasuhiko Ishikawa and Kazumi Wada

Department of Materials Engineering, The University of Tokyo, Tokyo 113-8654, Japan

ABSTRACT

Three-dimensional structures of microelectro-mechanical systems (MEMS)-based Ge waveguide on a Si beam were fabricated for dynamic tuning of the fundamental absorption edge of Ge by external stressing. The application of various amounts of external forces up to 1 GPa onto the Si beam shows clear red-shifts in the absorption edge of Ge waveguides on the Si beam by ~40 nm. This shift was reproduced by the deformation potential theory, considering that mode of propagation in the Ge waveguide. The wavelength tuning range obtained makes it possible to cover the whole C-band of optical communication, indicating it to be a promising approach to electro-absorption Ge modulators to get them to work with a broader wavelength range than previously reported.

KEYWORDS

modulators; electro-refractive; dynamic tuning, integrated photonics; MEMS

1. INTRODUCTION

Si photonics is an enabler of electronic and photonic integration on a Si CMOS (complementary metal oxide semiconductor) platform for cost-effective, bandwidth-abundant computation and communication on-chip networks. The current device challenge in Si photonics is the optical modulator to reduce the power consumption. Electro-refraction modulators (ERMs) such as Mach-Zehnder Interferometer (MZI) modulators [1,2] reported high power consumption in the order of ~pJ/bit because of the plasma effect that takes place to achieve the refractive index change of the long waveguide arm. Another type of ERM is ring resonator modulators [3,4], which do not require high power to operate but do require thermal management to match the resonance with the modulation wavelength. In contrast, electro-absorption modulators (EAMs) are essentially free from large power consumption because of the operation under reverse bias. The EAMs reported so far are based on the Franz-Keldysh effect of GeSi and Ge [5,6,7,8,9,10,11] and the quantum confined Stark effect of SiGe/Ge quantum wells [12,13]. The current challenge in EAMs is to expand their operation wavelength range, *i.e.*, currently only ~10 nm with a reasonable extinction ratio (~10 dB). Since the range is limited by the fundamental absorption edge, Ge alloying with Si [7] and stress-compressive Ge [11] can tune the operation wavelength of EAMs to 1550 nm. However, the range of the operation wavelength remained unchanged in these approaches.

The present paper has demonstrated dynamic tuning of the fundamental absorption edge of Ge waveguides in terms of applying stress in Ge waveguides on a Si beam (cantilever). The stress tuning concept was applied in the three-dimensional microelectro-mechanical systems (MEMS)-based beam structures and the bandgap shifts from applying stress were detected by the photoluminescence peak shifts [14,15,16,17,18]. To apply the concept to expand the range of the operation wavelength in terms of external stress, Ge waveguides are designed to be implemented on a Si waveguide on the Si beam. The devices have demonstrated ~40 nm red-shifts in the fundamental absorption edge of the Ge waveguides by dynamic external forces up to ~1 GPa on the Si beam. The shifts were consistent with the deformation potential calculations taking the propagation mode in the Ge waveguide into account. This prototype should be an enabler of the Ge EAMs with reconfigurable wavelength in the whole C-band in terms of external dynamic stressing.

2. DESIGN OF GE WAVEGUIDE TO TUNE THE FUNDAMENTAL ABSORPTION EDGE

To design the Ge waveguide on a MEMS-based Si beam, the structure should be optimized to meet both optical and mechanical requirements. In this chapter, the design is optimized in terms of controlling the fundamental absorption edge via stress, the structure of the Ge waveguide on the Si beam, and the mode of propagation.

2.1. Design of the Fundamental Absorption Edge and Stress

The dependence of the fundamental absorption edge of the Ge waveguide on one-dimensional stress along the <110> direction is calculated on the basis of deformation potential theory [18] as shown in Figure 1. The reason to choose the direction is related to Ge epigrowth and is discussed later. Built-in two-dimensional tensile stress is accumulated in the Ge epilayer grown on Si, and it splits the valence band into heavy hole (HH) and light hole (LH) even at no <110> stress [14]. Here, Γ-HH (Γ-LH) denotes the electron transition from the top of the HH (LH) band to the bottom of the Γ valley. It is clear that both Γ-HH and Γ-LH show "red-shifts" when one-dimensional tensile stress is applied to Ge. This indicates that the fundamental absorption edge of a Ge waveguide should be dynamically tuned depending on the amount of stress applied.

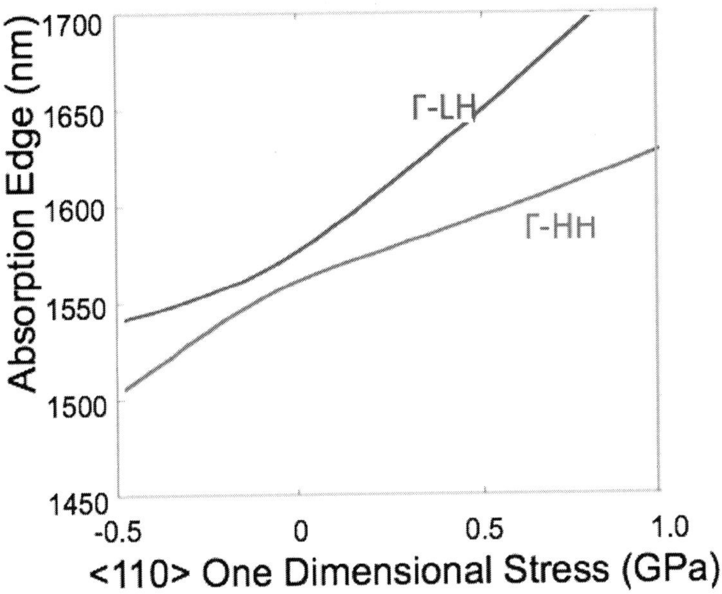

Figure 1. Fundamental absorption edge *vs.* <110> one-dimensional tensile stress of Ge. Ge has a built-in strain of 0.17% biaxial tensile due to growth cooling. This generates splitting of heavy hole (HH) and light hole (LH) even when no stress is applied. Applying one dimensional <110> tensile stress to the beam produces red-shifts in the absorption edge. The detailed calculations of the dispersion relation were reported elsewhere [11].

2.2. Design of Ge Waveguide on Si Beam Structure and Stress

The structure of the Ge waveguide on the Si beam and its effective stressing are described. Finite domain time difference (FDTD) simulations are used to design single mode Ge waveguides on the beam, as schematically shown in Figure 2: a 5-μm-long, 400-nm-thick and 400-nm-wide Ge waveguide on a Si rib waveguide with a 150 nm rib top on the Si on the insulator (SOI) wafer. The SOI wafer has a 250-nm-thick top Si layer on a 3-μm-thick buried oxide (BOX) layer. Here, the Si rib waveguide is employed to mechanically support the waveguide after removing the BOX layer under the Si beam. Employing the rib structure makes the bending radius of the Si waveguide large to minimize radiation losses. The 150-nm-high Si rib waveguide allows a bending radius of 10 μm which remains lower than 0.1 dB at the right angle bend, based on mode solver simulations. Based on these optimizations, our design of the Si beam to implement the Si rib waveguide should be 30 μm wide and 20 μm long. Such a large beam structure requires a specific design for effective stressing of the Ge waveguide. The neck structures as shown in Figure 2a are to enhance the stress to the Ge waveguide on the wide beam. The finite element method simulation of stress is performed to optimize the neck to be 5 μm wide and 5 μm long in shape and it indicates the amount of tensile stress to the Ge waveguide should be as high as 1 GPa when bent down to the base Si.

The finite element method (FEM) reveals the stress distribution in the Ge waveguide on the neck, as shown in Figure 2c. The stress should have a linear depth profile, and thus the top of the Ge waveguide has the highest stress (red) and the bottom has the lowest (green).

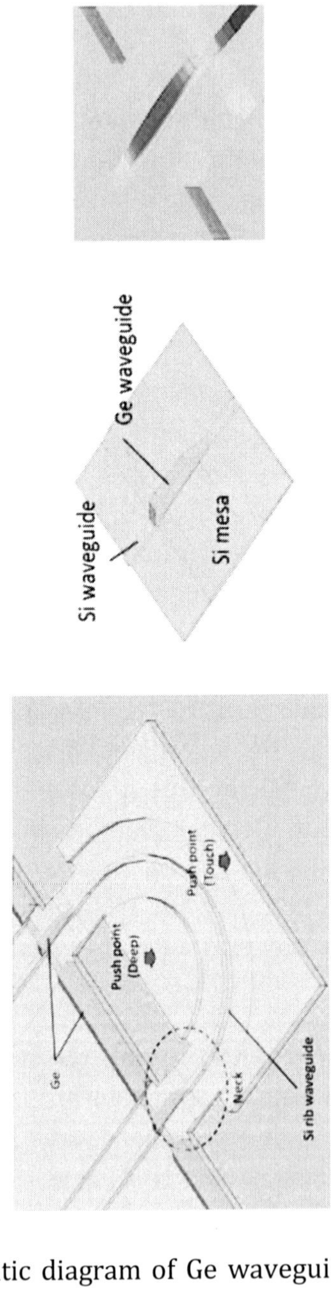

Figure 2. Schematic diagram of Ge waveguide on Si rib waveguide on Si beam. (**a**) The beam structure with two necks; (**b**) Magnified view near the neck region indicated by a dotted circle in (a), and (**c**) stress distribution by FEM.

The stresses at the typical point of the Ge waveguide are qualitatively shown in Table 1 by taking reference points such as "Ge top", a half-point of the thickness as "Ge middle", and the interface between Ge and Si as "Ge bottom". Here, an external force is applied at "Touch" and "Deep" until the beam touches down to the base Si. It has a peak near the waveguide direction. Here, the propagation mode would be the TE mode and the mode field is the highest near the Ge middle in steady state. Thus, the calculated amounts of stress at the Ge middle should be 0.2 and 1.3 GPa when applying stress to "Touch" and "Deep" and, needless to say, 0 GPa at "No stress". These stress amounts are taken at the peak point along the wavelength direction. The predicted fundamental absorption edges are calculated, using Figure 1, some of which are shown in Table 1: referencing the Ge middle point at "No stress", the red-shifts of Γ-HH should be 15 nm at "Touch" and 90 nm at "Deep", while those of Γ-LH should be 35 nm at "Touch" and 200 nm at "Deep".

2.3. Design of Propagation Mode in the Ge Waveguides

The coupling points of the Ge waveguide with the Si rib waveguide are shown in Figure 2 and Figure 3. The wedge-shaped Ge waveguide coupling with the Si rib waveguide should virtually generate a graded index structure to avoid unwanted scattering loss in terms of a sudden index change that the propagation mode experiences. It has so far been assumed that the propagation mode in the Ge waveguide should be in steady state. However, FDTD simulations of the mode propagation in the Ge and Si waveguide coupling indicate that the propagation mode is not in steady state but is transient in the 5-μm-long Ge waveguide, as shown in Figure 3. In other words, the mode bounces ups to the Ge middle and down to the Ge bottom or the Si rib underneath. Accordingly, the propagation mode should experience an in-depth profile of the stress field accumulated in the Ge waveguide, as shown in Figure 2c.

Table 1. Stress accumulated in the Ge waveguide.

Stress Application Point	Tensile Stress Amount (GPa) (Absorption Edges (nm) of Γ-HH and Γ-LH, Calculated from Figure 1)		
	Ge Top	Ge Middle	Ge Bottom
No stress	0 (1560, 1570)	0 (1560, 1570)	0 (1560, 1570)
Touch	0.3 (1580, 1620)	0.2 (1575, 1605)	0.04 (1565, 1582)
Deep	2.1 *	1.31.3 (1650, 1770)	0.17 (1575, 1605)

*: The ammount of stress 2.1 GPa may be too large to extrapolate it to the bandgap energies. Thus, there are no indications of corresponding wavelengths like the other cases in Table 1.

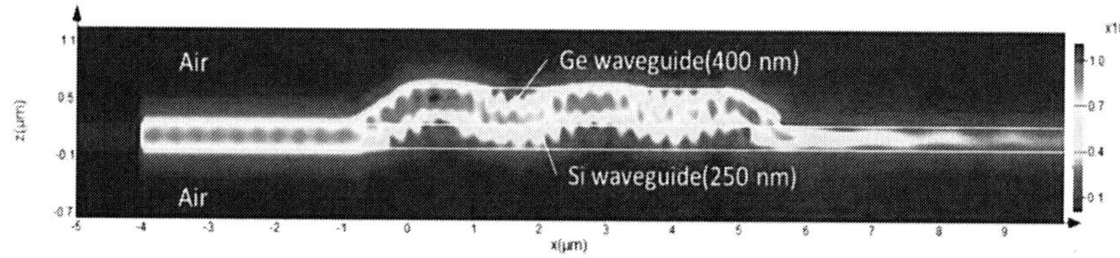

Figure 3. FDTD simulation of the propagation mode of 1550 nm in the wedge-shaped Ge waveguide on the Si rib waveguide.

Considering that the propagation mode is transient, the transmittance T of the mode through the Ge waveguide should be expressed by:

$$T = I_0 \left(1 - e^{-\iint \alpha(s(x,z)) \times L(x,z) dx dz}\right) \tag{1}$$

Here, I0 denotes the incoming power of the propagation mode into the Ge waveguide, α the absorption coefficient given in the Appendix, s(x,z) the stress in the Ge waveguide as a function of the length (x) and the height (z) of the waveguide, as shown in Figure 2c, and L(x, z) the trajectory of the propagating mode in the Ge waveguide, as shown in Figure 3. The rigorous solution of the transmittance has not been obtained from Equation (1), but the red-shift amount at the Ge bottom would be smaller than the prediction in Table 1. We will discuss on it later in Section 5.

3. FABRICATION OF GE WAVEGUIDE ON THE SI BEAM

Fabrication of the 20 × 30 μm beam ends up with a large removal of the surrounding BOX layer under the Si slab. A hydrofluoric acid (HF) solution is used to etch the BOX layer under the Si beam, which also removes the same amount of the BOX layer under the Si slab, referred to as the "undercut". Bending the beam with a large undercut defocuses stress through the whole structure on the undercut and stress on the Ge waveguide is reduced. To make the undercut small, the beam is designed with many small through-holes 0.5 μm in diameter and arrayed every 2.5 μm in a square lattice. FDTD simulations suggest that the propagation mode in the Si rib waveguide would couple to the top of the base Si wafer, when the beam touches down to the base Si, coupling in radiation mode. To isolate the propagation mode even at touching down, the BOX layer is partially left at the top of the base Si, and thus SiO_2 prevents the propagation mode from coupling to the base Si.

To achieve the wedge-shaped Ge waveguide edges, selective epitaxial growth (SEG) of Ge on Si has been used. The <311> faceted mesas on the sidewalls appear at the edge of the SEG Ge, which aligned to the {110} planes. The angle between the facet and (100) plane is 25.2°, which should be small enough to form graded index structures at the Ge waveguide, as shown in Figure 3. Thus, the stressing direction should also be in parallel to the <110> directions, as shown in Figure 1.

The Ge waveguides at the Si beam necks were devised on the top Si layer, using electron beam lithography and reactive ion etching. The 400-nm-thick Ge layer was first grown by an ultra-high-vacuum chemical vapor deposition method. A SiO_2layer was used as the SEG Ge mask to form (311) faceted wedge-shaped Ge waveguide edges. Reactive ion etching was performed by a Cl_2 and O_2 gas mixture plasma to fabricate a 400-nm-thick Ge waveguide and 150-nm-thick Si rib waveguide.

4. RESULTS AND ANALYSES

Figure 4 shows a typical SEM micrograph of the fabricated Si beam. The Si beam with two necks, the Si rib waveguide with a 10 μm bending radius on the beam, and two Ge waveguides on the Si rib waveguide on the necks are clearly shown, as designed. The white "dot" array on the beam is the through-holes to remove the BOX layer for fabrication of the free-standing beam and to maintain a thin SiO_2 layer under the beam for optical isolation.

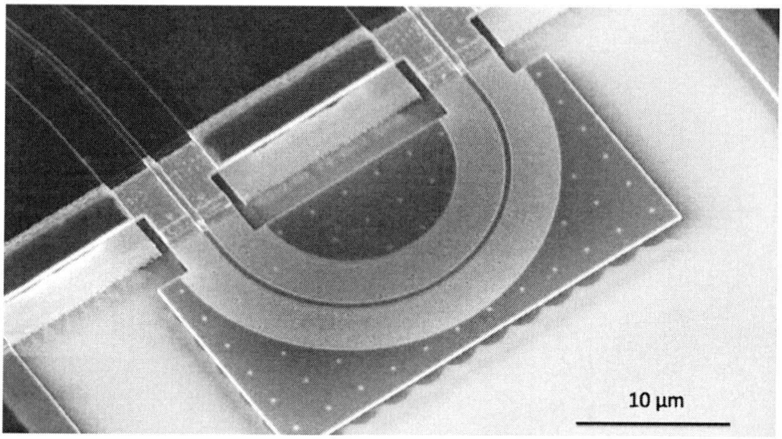

Figure 4. SEM micrograph of the Si beam with two necks. The curved Si rib waveguide and the Ge waveguide on each neck are clearly observed. The many small (white) dots observed on the Si beam are the through-holes to provide a HF solution under the beam.

Figure 5 shows transmission spectra of the Ge waveguide under no stress (green) applied to it other than that applied to the "Touch" (red) and "Deep" (blue) points, as defined in Figure 1a. The curve fittings are performed based on the calculation of the absorption coefficient, due to the Ge indirect bandgap that is calculated using one phonon model [19]. The details are described in the Appendix. The fittings well reproduce the experiment results, where the temperature-dependent constant at 291 K is used at 1900 cm^{-1} for the best fits shown in the Appendix. It is experimentally shown from the results that the red-shift by stressing is ~7 nm at "Touch", and

~52 nm at "Deep", assuming 2 × 5-μm-long Ge waveguides on the necks. Here, the beam was bent by applying an external force to touch down to the base Si to get these red-shifts. The absorption curve at "No stress" was used as the reference.

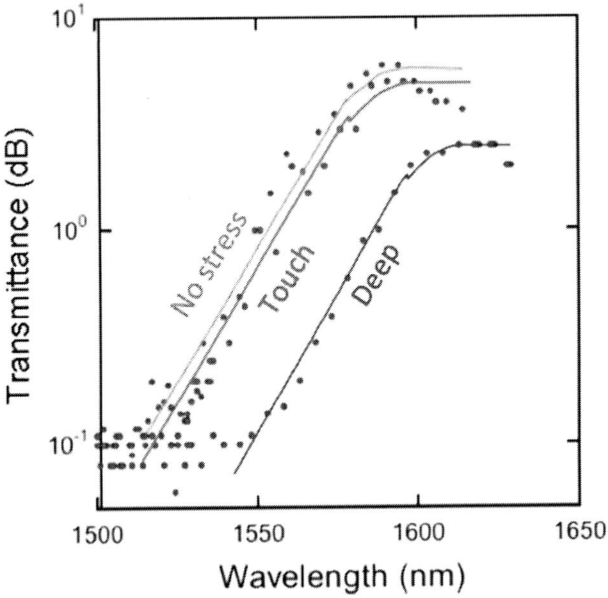

Figure 5. Transmission spectra of the Ge waveguide on the Si beam neck. The green denotes "No stress", the red "Touch" and the blue "Deep", as in Figure 1a. A small red-shift is detected when stress is applied to the "Touch" point, while larger red-shift is detected when stress is applied to the "Deep" point. The solid lines are the fitting results using the one photon model calculation described in the Appendix.

These observations indicate that the absorption edge at the propagation mode is to experience is Γ-HH as shown in Figure 1. It may be fair to exclude Γ-LH since the red-shifts are too large (35 nm and 200 nm at the "Touch" and "Deep" points). The red-shifts by Γ-

HH as shown in Table 1 are 15 nm and 90 nm at stressing the "Touch" and "Deep" points, which are twice larger compared with the observations. The discrepancy is due to the fact that the propagation mode is transient and the stress profiles in the depth and along the waveguide are inhomogeneous. It is rather hard to predict how the transmission will behave. We will come back to this point in the discussion. Here, it is concluded that applying dynamic stress to the Ge waveguide on the Si beam could precisely tune its fundamental absorption edge at any wavelength in the range of 40 nm. In other words, Ge EAMs, if fabricated on the beam, will cover the entire C-band range (35 nm) in the optical communication wavelength by dynamic stressing at the "Deep" point.

5. DISCUSSION

To understand the discrepancy between experimental transmittances in Figure 5 and theoretical predictions made by Equation (1), first we should assume that the trajectory of the propagating mode in the Ge waveguide does not change upon the strength of stressing. In other words, it is always the same as the one shown in Figure 3. We can define L(x, z) in Equation (1) as a sinusoidal curve; the period along the *x* direction is 2 μm and the intensity is half of the Ge waveguide height (H) with z = 0 at the Ge and Si interface. We further assume that the stress along the Ge waveguide direction is uniform although actual distribution exists, as shown in Figure 2c. This leads us to express Equation (1) as:

$$T = I_0 \left(1 - \int dx e^{-\alpha(s_{max}(\sin x)) \times \mathrm{H} \sin x} \right) \tag{2}$$

Here, smax is the maximum stress the mode field experiences depending on the stress point. The point should be the Ge middle. A simple calculation using Equation (2) showed us that the red-shift of the fundamental absorption edge becomes 7.5 nm at "Touch" and 45 nm at “Deep”. These agree with the observed red-shifts in Figure 5, *i.e.*, ~7 nm at “Touch” and 52 nm at “Deep”, despite the fact that the non-uniformity of stress along the Ge waveguide direction as shown in Figure 2c is ignored in this calculation. This leads us to conclude that the discrepancy between the predicted red-shifts in Table 1 and the observed ones in Figure 5 should be induced by the transient propagation mode in the Ge waveguides.

The transmittance results in Figure 5 show clear reductions at the wavelength range beyond 1600 nm, despite the fact that the theoretical curves do not show such decreases. The difference can also be explained by the transient propagation mode in Figure 3. The propagation mode smoothly couples down to the Si rib waveguide as shown in Figure 5. It should be noted that the wavelength is 1550 nm in the FDTD simulations. The propagation mode should hit the (311) Ge waveguide wedge when the wavelength becomes different from 1550 nm. This is the reason why the transmittance is reduced beyond 1600 nm. Such a reduction should have happened to the wavelength range around 1500 nm. However, the intensities of the transmittance are below the noise level of these measurements and thus it is not detectable.

Finally, it should be proposed that the stress tuning of the Ge waveguide could be done using an electrostatic force applying a voltage across the beam and the base Si wafer [20]. By fabricating an electrode at “Deep” on the Si beam and applying a voltage across the electrode and the base Si to touch the “Deep” point down to the base Si, the 52 nm red-shift should be reproduced. In this configuration, the SiO_2 layer intentionally left on the base Si should work for isolating the Si beam not only optically but also electrically from the Si wafer.

6. CONCLUSIONS

The present paper has proposed a way to dynamically tune the fundamental absorption edge of Ge waveguides by applying tensile stress to them and it described a prototype Ge waveguide on a Si beam structure. The three-dimensional structures of microelectro-mechanical systems (MEMS)-based Ge waveguide on a Si beam are optimized in terms of stress, mechanical stability, and propagation mode and then fabricated. It has been demonstrated that the absorption edge of two Ge waveguides on the Si beam can be dynamically tuned to show the red-shift in the wavelength range of ~52 nm by applying an external force of 1 GPa to the "Deep" point of the Si beam. The experimentally observed red-shifts are well explained, considering the transient mode of propagation in the Ge waveguides. The presented dynamic tuning method should enable the reconfiguration of the operation wavelength range of Ge electro-absorption modulators in the entire C-band. We propose that an electro-static force can be applied to achieve on-chip stress tuning devices in practical applications.

ACKNOWLEDGMENTS

This research project was partly supported by the Ministry of Education, Culture, Sports, Science and Technology (MEXT), Japan (Nanotechnology Platform, Project No. 12024046).

AUTHOR CONTRIBUTIONS

M.H. and L.M.N. designed and performed the experiments; H.F. performed electro-static tuning experiment; Y.I. and K.W. co-supervised the studies.

CONFLICTS OF INTEREST

The authors declare no conflict of interest.

APPENDIX

In the one phonon model [19], the indirect absorption has the quadratic dependence in the energy difference between the photon energy on one hand and the band gap energy plus or minus phonon energy.

$$\alpha = A\left[\frac{1}{1-e^{-\Theta/T}}\left(\frac{\hbar\omega - E_{G.Ind} - k\Theta}{h\nu}\right)^2 + \frac{1}{e^{\Theta/T}-1}\left(\frac{\hbar\omega - E_{G.Ind} + k\Theta}{h\nu}\right)^2\right] \quad \text{(A1)}$$

Here, A denotes a temperature-dependent constant (cm^{-1}), kΘ the phonon energy, ħω the energy of the photon, EG.Indthe indirect band gap (0.664 eV) and T the temperature in Kelvins. For simplicity, we use the longitudinal acoustic phonon of Ge with the energy of 27.7 meV. The built-in stress in the Ge epilayer on Si splits the valence band into the *LH* and the *HH*band, as in Figure 1. The contribution of each indirect band to the total absorption, α(total), depends on the effective masses and can be formulated as follows:

$$\alpha\,(total) = k_1\alpha\,(LH) + k_2\alpha\,(HH) \quad \text{(A2)}$$

where $k_1 = \frac{m_{r,lh}{}^{3/2}}{m_{r,lh}{}^{3/2} + m_{r,hh}{}^{3/2}}$, $k_2 = \frac{m_{r,hh}{}^{3/2}}{m_{r,lh}{}^{3/2} + m_{r,hh}{}^{3/2}}$, $m_{r,lh} = \frac{m_{lh}\, m_{e,L}}{m_{lh} - m_{e,L}}$, and $m_{r,hh} = \frac{m_{hh}\, m_{e,L}}{m_{hh} + m_{e,L}}$. The electron mass at *L*valley me,L is 0.22 m_o. Using Equations (A1) and (A2), we obtained Figure 4 to fit the experimental data.

REFERENCES

1. Green, W.M.; Rooks, M.J.; Sekaric, L.; Vlasov, Y.A. Ultra-compact, low RF power, 10 Gb/s silicon Mach-Zehnder modulator. *Opt. Express* **2007**, *15*, 17106–17113.

2. Reed, G.T.; Mashanovich, G.; Gardes, F.Y.; Thomson, D.J. Silicon optical modulators. *Nat. Photonics* **2010**, *4*, 518–526.

3. Xu, Q.; Schmidt, B.; Pradhan, S.; Lipson, M. Micrometre-scale silicon electro-optic modulator. *Nature* **2005**, *435*, 325–327.

4. Watts, M.R.; Zortman, W.A.; Trotter, D.C.; Young, R.W.; Lentine, A.L. Vertical Junction silicon microdisk modulators and switches. *Opt. Express* **2011**, *22*, 21989–22003.

5. Jongthammanurak, S.; Liu, J.F.; Wada, K.; Cannon, D.D.; Danielson, D.T.; Pan, D.; Kimerling, L.C.; Michel, J. Large Electro-optic Effect in Tensile Strained Ge-on-Si Films. *Appl. Phys. Lett.* **2006**, *89*.

6. Liu, J.F.; Pan, D.; Jongthammanurak, S.; Wada, K.; Kimerling, L.C.; Michel, J. Design of monolithically integrated GeSi electro-absorption modulators and photodetectors on a SOI platform. *Opt. Express* **2007**, *15*, 623–628.

7. Liu, J.F.; Beals, M.; Pomerene, A.; Bernardis, S.; Sun, R.; Cheng, J.; Kimerling, L.C.; Michel, J. Waveguide integrated, ultra-low energy GeSi electro-absorption modulators. *Nat. Photonics* **2008**, *2*, 433–437.

8. Lim, A.E.; Liow, T.; Qing, F.; Duan, N.; Ding, L.; Yu, M.; Lo, G.; Kwong, D.-L. Novel evanescent coupled germanium electro absorption modulator featuring monolithic integration with germanium PIN photodetector. *Opt. Express* **2010**,*19*, 5040–5046.

9. Feng, N.N.; Feng, D.; Liao, S.; Wang, X.; Dong, P.; Liang, H.; Kung, C.C.; Qian, W.; Fong, J.; Shafiiha, R.; *et al.* 30 GHz Ge electro-

absorption modulator integrated with 3 μm silicon-on-insulator waveguide. *Opt. Express* **2011**, *19*, 7062–7067.

10. Feng, D.; Liao, S.; Liang, H.; Fong, J.; Bijlani, B.; Shafiiha, R.; Luff, B.J.; Luo, Y.; Cunningham, J.; Krishnamoorthy, A.V.;*et al.* High speed GeSi EA modulator at 1550 nm. *Opt. Express* **2012**, *20*, 22224–22232.

11. Nguyen, L.M.; Kuroyanagi, R.; Tsuchizawa, T.; Ishikawa, Y.; Yamada, K.; Wada, K. Stress tuning of the fundamental absorption edge of pure germanium waveguides. *Opt. Express* **2015**, *23*, 18487–18492.

12. Kuo, Y.-H.; Lee, Y.K.; Ge, Y.; Ren, S.; Roth, J.E.; Kamins, T.I.; Miller, D.A.B.; Harris, J.S. Quantum-confined stark effect in Ge-SiGe quantum wells on Si for optical modulators. *IEEE J. Sel. Top. Quantum Electron.* **2006**, *12*, 1503–1513.

13. Chaisakul, P.; Marris-Morini, D.; Roulfed, N.-S.; Frigerio, J.; Charastina, D.; Coudevylle, J.-P.; Roux, X.L.; Edmong, S.; Isella, G.; Vivien, L. Recent Progress in GeSi electro-absorption modulators. *Sci. Technol. Adv. Mater.* **2014**, *15*, 14601–14608.

14. Ishikawa, Y.; Wada, K.; Cannon, D.D.; Liu, J.; Luan, H.-C.; Kimerling, L.C. Strain-induced band gap shrinkage in Ge grown on Si substrate. *Appl. Phys. Lett.* **2003**, *82*.

15. Yoshimoto, K.; Suzuki, R.; Ishikawa, Y.; Wada, K. Bandgap control using strained beam structures for Si photonic devices. *Opt. Express* **2010**, *18*, 26492–26498.

16. Cai, J.; Ishikawa, Y.; Wada, K. Strain induced bandgap and refractive index variation of silicon. *Opt. Express* **2013**, *21*, 7162–7170.

17. Horie, Y.; Decostered, L.; Suzuki, R.; Ishikawa, Y.; Wada, K. Emission wavelength tuning by mechanical stressing of GaAs/Ge/Si microbeams. *Opt. Express* **2011**, *19*, 15732–15738.

18. Wada, K.; Yoshimoto, K.; Horie, Y.; Cai, J.; Lim, P.H.; Fukuda, H.; Suzuki, R.; Ishikawa, Y. Srtrained Ge for Si-Based Integrated Photonics. In *Photonics and Electronics with Ge*; Wada, K., Kimerling, L.C., Eds.; Wiley-VHC: Weinheim, Germany, 2015; p. 219.

19. Macfarlane, G.G.; Roberts, V. Infrared Absorption of Germanium near the Lattice Edge. *Phys. Rev.* **1955**, *97*.

20. Fukuda, H. A Study of Silicon Microphotonics for Telecommunication Applications. Ph.D. Thesis, the University of Tokyo, Tokyo, Japan, 2014.

3

HIGH-*Q* DEFECT-FREE 2D PHOTONIC CRYSTAL CAVITY FROM RANDOM LOCALISED DISORDER

Kelvin Chung [1,*], Timothy J. Karle [1], Ranjith Rajasekharan [1], C. Martijn de Sterke [2] and Snjezana Tomljenovic-Hanic [1]

[1] School of Physics, The University of Melbourne, Parkville, Victoria 3010, Australia

[2] School of Physics, The University of Sydney, New South Wales 2006, Australia

ABSTRACT

We propose a high-*Q* photonic crystal cavity formed by introducing random disorder to the central region of an otherwise defect-free photonic crystal slab (PhC). Three-dimensional finite-difference time-domain simulations determine the frequency, quality factor, *Q*, and modal volume, *V*, of the localized modes formed by the disorder. Relatively large Purcell factors of 500–800 are calculated for these cavities, which can be achieved for a large range of degrees of disorders.

KEYWORDS

defect-free; 2D photonic crystal; positional disorder

1. INTRODUCTION

Photonic crystals (PhCs) have been widely investigated because they can exhibit a photonic bandgap, *i.e.*, a frequency region in which electromagnetic waves cannot propagate within the structure. This is achieved through the periodic variation of the dielectric constant of the structure, usually implemented through the fabrication of a periodic array of holes in a thin slab of high-index material [1]. In such 2D PhCs the periodic array holes leads to Bragg reflection in the plane of the slab, which, upon appropriate design, leads to the formation of a photonic bandgap. Meanwhile, the light is confined vertically in the slab by the total internal reflection. The presence of a photonic bandgap in PhCs gives rise to a variety of potential applications, for example they can act as filters [2], optical switches, as promising components for photonic integrated circuits [3], and as defect laser cavities [4]. PhC-based cavities offer compact and efficient optical confinement with high quality factors (Q) [5] and small modal volumes (V), which is advantageous for cavity QED experiments, for example using 2D geometries which can be coupled to quantum emitters [6,7]. In particular, the high-Q and small V offers the enhancement of the spontaneous emission rate of a coupled emitter-cavity system [8]. Various emitters that have been successfully coupled to PhC cavities include quantum dots [9,10,11,12] and colour centres [13,14,15]. PhC cavities and the associated modal confinement can be achieved by introducing a defect in the periodic lattice such as removing a hole from a 2D lattice of airholes, which leads to a localized state in the photonic bandgap. An alternative is a defect-free PhC cavity, *i.e.*, a cavity in which none of the air holes is removed from the periodic lattice [16,17,18,19]. Rather, such cavities result in PhCs made of photosensitive chalcogenide glass [20], by the introduction of nanodiamonds into airholes [21], and in PhCs with airholes with graded radii [16,22]. The effect of fabrication-related positional disorder on defect-free PhC cavities has been reported [17,18,19], however these studies all consider position disorder of the airholes in the entire PhC.

The fabrication of 2D PhC cavities for optical wavelength generally requires high-precision lithography techniques such as electron beam or X-ray lithography [23,24], since the geometrical features of PhCs designed are on the sub-wavelength scale. Inevitably, the inability to accurately fabricate features with a size which is on the order of a visible wavelength leads to imperfection. Structural characteristics which are affected during fabrication include surface roughness, sidewall angle, and the size and position of holes [25]. This has attracted much research to determine the tolerance of structures to variation of physical parameters which perturb the photonic bandgap [19,26,27,28]. For PhC cavities, the parameters of interest are the resonant frequency, the quality factor, and modal volume.

In this paper, we show that high-Q cavities can be achieved by introducing local random disorder of the position of the airholes in the region around the centre of the 2D PhC. In addition to high-Q and small V, we show that such cavities can exhibit a large Purcell enhancement, which is associated with strong coupling between matter-photon interactions. The randomness of the cavity formation leads to a resonant wavelength and field distribution which cannot be predicted, which may be alleviated by the use of fluorescent molecules with slightly different resonant frequencies distributed throughout the proximity of the cavity. Section 2 outlines the theory and method, Section 3 contains the results and associated discussion.

2. THEORY AND METHOD

We consider a 2D silicon (n_{Si} = 3.4) PhC slab with a hexagonal array of cylindrical air holes which extends for 36 periods both in the *x*- and *y*-directions (in-plane). The structure has a periodicity of a = 0.449 μm, slab thickness of $h = 0.6a$, and the radius of the cylindrical airholes is $R = 0.25a$. The plane positions of the 19 innermost airholes are all randomly perturbed. A schematic of the structure is shown in Figure 1a.

The perturbed positions of the airholes is defined by:

$$\Delta x_i = x_i + \delta f_{\mathrm{ran}}^i \quad (1)$$

where x_i is the unperturbed position of the *i*th airhole in either of the in-plane coordinates (see Figure 1a) of the random region, δ is the amplitude of randomness which controls the minimum and maximum value of the shift, and f_{ran} returns a random uniformly distributed number that lies within the range [−0.5,0.5]. Note, results shown below refer to a single realisation of f_{ran} only. We have also considered different f_{ran} functions, *i.e.*, ensemble behaviour, and the same trends in frequencies, total *Q*s, modal volume and mode profiles were also observed for each instance f_{ran}.

The randomness generates regions of higher and lower refractive index with respect to the unperturbed PhC which manifests in a variation of the global minimum of $U_l(\delta)$ for Figure 1b, where U_l is the local potential well induced that traps the cavity mode induced by the disorder. The variation is not uniform; it is dependent on the position of the airholes which can be either negative or positive. The modes shown in Figure 1b exist in regions with the overall increased refractive index, which redistributes modes locally deeper into the bandgap.

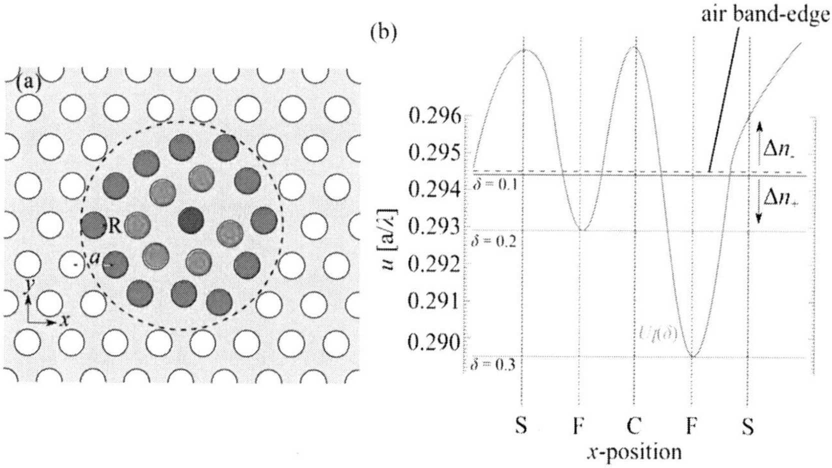

Figure 1. (**a**) Schematic of the hexagonal array 2D silicon (grey) photonic crystal (PhC) slab with the dashed circle showing the boundary inside which the position of the cylindrical airholes are randomly perturbed. The origin, *i.e.*, (*x*,*y*) = (0,0), is measured from the centre of the unperturbed blue airhole. Note: the centre airhole is perturbed in the figure; (**b**) Normalised frequency *u* of high-*Q* modes found for various disorder amplitudes, δ, (solid lines) relative to the air-band edge of the cavity. An arbitrarily shaped local potential well $U_l(\delta)$ created from the local refractive index distribution governed by its dependence on δ and a random function (see Equation (1)). The positions of the airholes are indicated by the colours blue (C—centre), green (F—first ring) and red (S—second ring) as a function of the *x*-position of the airholes. In some regions of the 2D PhC slab, the local refractive index either increases (below the band gap), Δn_+, or decreases (above the band gap), Δn_-, depending on the random positions of the airholes. In cases where the refractive index is increased, it is possible to obtain localised states inside the bandgap as indicated by the modes with disorder amplitudes of δ = 0.1, 0.2 and 0.3.

3D finite-difference time-domain (using RSoft's FDTD implementation called FullWAVE, version 6.2) is used to simulate the modes of the disordered PhC. The computational domain window is halved by imposing field symmetry properties in the out-of-plane direction about the centre of the slab. Satisfactory convergence is obtained by using 28 points per period. At the computational domain boundaries, perfectly matched layers (PMLs) for width and height are $2a$ and $4h$, respectively.

Cavity modes can be coupled to emitters enhancing their spontaneous emission rate. The degree of enhancement is given by the Purcell factor F, is given by [29].

$$F = \frac{3}{4\pi^2}\left(\frac{\lambda}{n}\right)^3 \frac{Q}{V} \quad (2)$$

where Q is the total quality factor, λ is the resonant wavelength, and n is the refractive index. Q depends on the partial Qs that occur in-plane $Q_{||}$ and out-of-plane $Q_{\perp}$ with respect to the structure's orientation (in-plane coordinates: xy-plane; out-of-plane direction: z-axis) and are related via $1/Q = 1/Q_{||} + 1/Q_{\perp}$. The in-plane term is further related to the partial Qs in each respective Cartesian direction: $1/Q_{||} = 1/Q_x + 1/Q_y$ [30].

3. RESULTS AND DISCUSSION

The random positions of the holes causes backscattering which yield non-propagating Bloch waves which localises the electric field [12]. Figure 1b shows that by introducing positional disorder to the cavity, a resonant mode can enter the photonic band gap (PBG). The normalised frequencies of the cavity modes with high-Q are $u_{\delta=0.1}$ = 0.295, $u_{\delta=0.2}$ = 0.293 and$u_{\delta=0.3}$ = 0.290. Increasing δ causes the modes to shift down in frequency away from the air band-edge and towards the centre of the PBG, which in turn corresponds to a mode which is more strongly confined in-plane. These modes are localised due to a local potential well being created and its functional form depends on δ and on the realization of the random process. Our calculations recover the frequency, the total and partial *Q*s and *V* of the disordered cavity modes. These results are given in the next section.

3.1. Total, Partial Qs and Purcell Effect

The total and partial *Q*s calculated for the cavity modes for increasing δ are shown in Table 1.

From Table 1, we note that the in-plane *Q*s increase with δ and whereas the out-of-plane *Q*sdecrease. This can be understood as follows: When the deviation from perfect periodicity increases more *k*-components enter the light cone leading to an increase in out-of-plane radiation losses. In contrast, as δ increases, the defect state goes deeper into the bandgap, thus improving the in-plane confinement and hence leading to an increase in $Q_{||}$.

In Figure 2, we show the calculated partial in-plane $Q_{||}$ (crosses) and out-of-plane $Q_{\perp}$ (squares) as a function of δ. It can be seen that both partial *Q*s behave very differently, $Q_{||}$ increases with random disorder, and conversely, $Q_{\perp}$ decreases. The modal volumes were also calculated for the range of δ = 0.1 and 0.3 cavities, where they were found to vary between $2.03(\lambda/n)^3$ and $0.93(\lambda/n)^3$ respectively (see Table 1). As mentioned in Section 2 the same trends in total and partial *Q*s were also obtained for different f_{ran}. We have tested four different random functions.

3.2. Mode Profiles

The major electric field components E_x and E_y, and their Fourier transforms, of the in-plane coordinates for the disorder amplitudes δ = 0.1, 0.2 and 0.3 are shown in Figure 3. Figure 3a,c show the changes in-plane confinement for each respective mode given their associated disorder amplitude and it can be seen that the confinement around the central region of cavity improves as the disorder increases, subsequently V decreases. This is confirmed quantitatively since $Q_{||}$ from Figure 2varies from 1.90×10^4 to 3.82×10^5 for increasing disorder. Another indicator is the k-space distribution of Figure 3b,d which shows that the fraction of the field that falls within the leaky light-cone is higher for larger δ, confirming the behaviour seen in Figure 2 [30].

Table 1. Total, partial Qs ($Q_\perp$,$Q_{||}$), V and F of the cavity modes as a function of disorder amplitude δ. The modal volume for δ = 0.05 could not be determined because the computational domain was too large.

δ	$Q_\perp$ (×10⁴)	$Q_{\parallel}$ (×10⁴)	Q (×10⁴)	V $(\lambda/n)^3$	F
0.05	20.2	0.45	0.44	–	–
0.1	12	1.9	1.65	2.03	618
0.15	3.77	2.5	1.5	1.6	712
0.2	1.5	10.3	1.21	1.20	766
0.25	1.1	23.8	1.1	1.02	820
0.3	0.66	38.2	0.65	0.93	531

4. CONCLUSIONS

We have shown through random local positional disorder that high-Q 2D PhC cavities can be achieved with significantly high Purcell enhancement up to F = 820 with a random disorder amplitude of δ = 0.25. This enhancement is approximately 55 times greater compared to Anderson-localized modes in a photonic crystal waveguide [31]. We have shown that a defect-free locally disordered 2D PhC slab is a promising geometry for large Purcell factors. This random disorder represents a fabrication disadvantage that can be directly addressed since cavity modes always exist below the air-band edge. Exploring this parameter space allows this seemingly fabrication disadvantage to become an advantage because modes will always be pulled into the bandgap and a high-*Q* can be achieved.

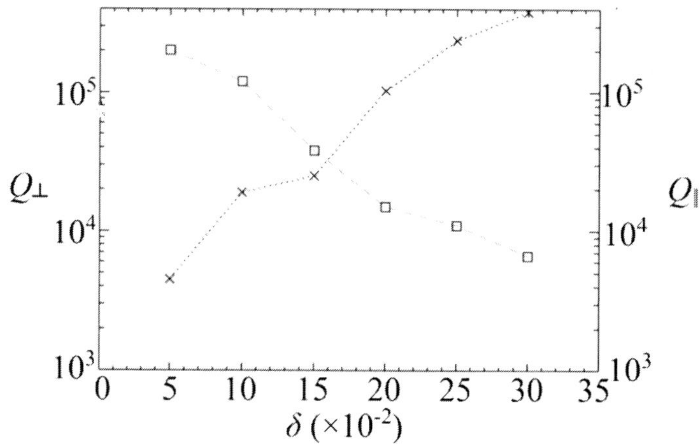

Figure 2. In-plane (crosses) and out-of-plane (square) partial *Q*s of the cavity as a function of disorder amplitude δ.

ACKNOWLEDGMENTS

K.C. would like to acknowledge the receipt of the Melbourne Research Scholarship awarded by The University of Melbourne. R.R. acknowledges McKenzie Fellowship Scheme awarded by The University of Melbourne. S.T.-H. is supported by the ARC Australian Research Fellowship (project number DP1096288). This research was also undertaken with the assistance of resources provided at the NCI National Facility systems at The Australian National University through the National Computational Merit Allocation Scheme supported by the Australian Government.

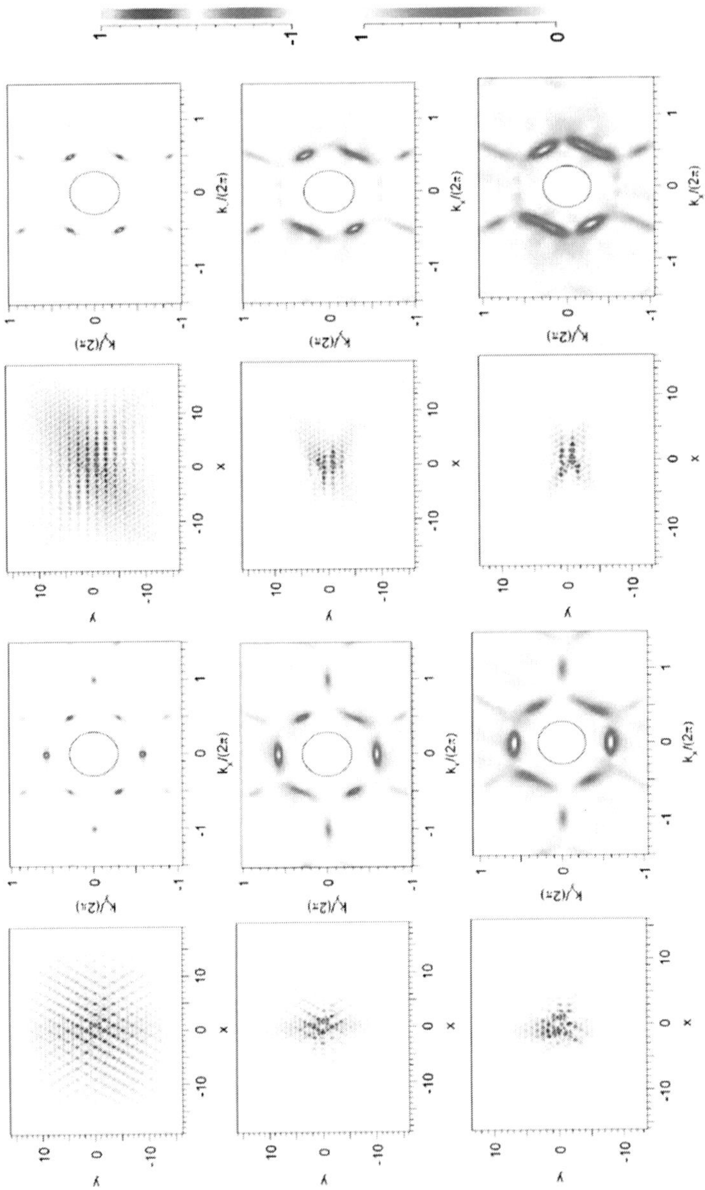

Figure 3. The major electric field components in the middle of the slab for (**a**) E_x and its associated; (**b**) Fourier transform. Similarly; the (**c**) E_y major field component; and its (**d**) Fourier transform (circle represents the light). The top, middle and bottom rows correspond to disorder amplitudes δ = 0.1, 0.2 and 0.3 respectively.

AUTHOR CONTRIBUTIONS

The conceptual idea was developed by S.T.-H., further discussion and clarification with C.M.dS. Simulations were conducted by K.C., T.J.K., S.T.-H. and R.R. The writing, editing and discussion of the paper was conducted by all authors.

CONFLICTS OF INTEREST

The authors declare no conflict of interest.

REFERENCES

1. Krauss, T.F.; Rue, R.M.D.L.; Brand, S. Two-dimensional photonic-bandgap structures operating at near-infrared wavelengths. *Nature* **1996**, *383*, 699–702.

2. Noda, S.; Chutinan, A.; Imada, M. Trapping and emission of photons by a single defect in a photonic bandgap structure. *Nature* **2000**, *407*, 608–610.

3. Intonti, F.; Vignolini, S.; Turck, V.; Colocci, M.; Bettotti, P.; Pavesi, L.; Schweizer, S.L.; Wehrspohn, R.; Wiersma, D. Rewritable photonic circuits. *Appl. Phys. Lett.* **2006**, *89*, 211117:1–211117:3.

4. Painter, O.; Lee, R.; Scherer, A.; Yariv, A.; O'Brien, J.; Dapkus, P.; Kim, I. Two-dimensional photonic band-gap defect mode laser. *Science* **1999**, *284*, 1819–1821.

5. Akahane, Y.; Asano, T.; Song, B.-S.; Noda, S. High-*Q* photonic nanocavity in a two-dimensional photonic crystal.*Nature* **2003**, *425*, 944–947.

6. Vučković, J.; Lončar, M.; Mabuchi, H.; Scherer, A. Design of photonic crystal microcavities for cavity QED. *Phys. Rev. E* **2001**, *65*, 016608:1–016608:11.

7. Kiraz, A.; Reese, C.; Gayral, B.; Zhang, L.; Schoenfeld, W.; Gerardot, B.; Petroff, P.; Hu, E.; Imamoglu, A. Cavity-quantum electrodynamics with quantum dots. *J. Opt. B Quantum Semiclass. Opt.* **2003**, *5*.

8. Aharonovich, I.; Greentree, A.D.; Prawer, S. Diamond photonics. *Nat. Photon.* **2011**, *5*, 397–405.

9. Yoshie, T.; Scherer, A.; Hendrickson, J.; Khitrova, G.; Gibbs, H.; Rupper, G.; Ell, C.; Shchekin, O.; Deppe, D. Vacuum rabi splitting with a single quantum dot in a photonic crystal nanocavity. *Nature* **2004**, *432*, 200–203.

10. Englund, D.; Faraon, A.; Fushman, I.; Stoltz, N.; Petroff, P.; Vučković, J. Controlling cavity reflectivity with a single quantum dot. *Nature* **2007**, *450*, 857–861.

11. Hennessy, K.; Badolato, A.; Winger, M.; Gerace, D.; Atatüre, M.; Gulde, S.; Fält, S.; Hu, E.L.; Imamoğlu, A. Quantum nature of a strongly coupled single quantum dot-cavity system. *Nature* **2007**, *445*, 896–899.

12. Yang, J.; Heo, J.; Zhu, T.; Xu, J.; Topolancik, J.; Vollmer, F.; Ilic, R.; Bhattacharya, P. Enhanced photoluminescence from embedded pbse colloidal quantum dots in silicon-based random photonic crystal microcavities. *Appl. Phys. Lett.* **2008**,*92*, 261110:1–261110:3.

13. Faraon, A.; Santori, C.; Huang, Z.; Acosta, V.M.; Beausoleil, R.G. Coupling of nitrogen-vacancy centers to photonic crystal cavities in monocrystalline diamond. *Phys. Rev. Lett.* **2012**, *109*.

14. Englund, D.; Shields, B.; Rivoire, K.; Hatami, F.; Vučković, J.; Park, H.; Lukin, M.D. Deterministic coupling of a single nitrogen

vacancy center to a photonic crystal cavity. *Nano Lett.* **2010**, *10*, 3922–3926.

15. Wolters, J.; Schell, A.W.; Kewes, G.; Nüsse, N.; Schoengen, M.; Döscher, H.; Hannappel, T.; Löchel, B.; Barth, M.; Benson, O. Enhancement of the zero phonon line emission from a single nitrogen vacancy center in a nanodiamond via coupling to a photonic crystal cavity. *Appl. Phys. Lett.* **2010**, *97*, 141108:1–141108:3.

16. Srinivasan, K.; Barclay, P.; Painter, O. Fabrication-tolerant high quality factor photonic crystal microcavities. *Opt. Express* **2004**, *12*, 1458–1463.

17. Moon, S.-K.; Yang, J.-K. Numerical study of the photonic-bandgap effect in two-dimensional slab photonic structures with long-range order. *J. Opt.* **2013**, *15*, 075704:1–075704:7.

18. Molinari, D.; Fratalocchi, A. Route to strong localization of light: The role of disorder. *Opt. Express* **2012**, *20*, 18156–18164.

19. Li, Z.-Y.; Zhang, X.; Zhang, Z.-Q. Disordered photonic crystals understood by a perturbation formalism. *Phys. Rev. B***2000**, *61*.

20. Tomljenovic-Hanic, S.; de Sterke, C.M. Design of ultrahigh-*Q* photoinduced cavities in defect-free photonic crystal slabs. *Opt. Express* **2010**, *18*, 21397–21403.

21. Tomljenovic-Hanic, S.; Greentree, A.D.; Gibson, B.C.; Karle, T.J.; Prawer, S. Nanodiamond induced high-*Q* resonances in defect-free photonic crystal slabs. *Opt. Express* **2011**, *19*, 22219–22226.

22. Srinivasan, K.; Barclay, P.E.; Painter, O.; Chen, J.; Cho, A.Y.; Gmachl, C. Experimental demonstration of a high quality factor photonic crystal microcavity. *Appl. Phys. Lett.* **2003**, *83*, 1915–1917.

23. Feiertag, G.; Ehrfeld, W.; Freimuth, H.; Kolle, H.; Lehr, H.; Schmidt, M.; Sigalas, M.; Soukoulis, C.; Kiriakidis, G.; Pedersen, T. Fabrication of photonic crystals by deep X-ray lithography. *Appl. Phys. Lett.* **1997**, *71*, 1441–1443.

24. Awazu, K.; Wang, X.; Fujimaki, M.; Kuriyama, T.; Sai, A.; Ohki, Y.; Imai, H. Fabrication of two- and three-dimensional photonic crystals of titania with submicrometer resolution by deep X-ray lithography. *J. Vac. Sci. Technol. B* **2005**, *23*, 934–939.

25. Khankhoje, U.; Kim, S.-H.; Richards, B.; Hendrickson, J.; Sweet, J.; Olitzky, J.; Khitrova, G.; Gibbs, H.; Scherer, A. Modelling and fabrication of GaAs photonic-crystal cavities for cavity quantum electrodynamics. *Nanotechnology***2010**, *21*, 065202:1–065202:7.

26. Fan, S.; Villeneuve, P.R.; Joannopoulos, J. Theoretical investigation of fabrication—Related disorder on the properties of photonic crystals. *J. Appl. Phys.* **1995**, *78*, 1415–1418.

27. Chutinan, A.; Noda, S. Effects of structural fluctuations on the photonic bandgap during fabrication of a photonic crystal. *J. Opt. Soc. Am. B* **1999**, *16*, 240–244.

28. Zhu, Z.; Ye, W.; Ji, J.; Yuan, X.; Zen, C. Influence of random errors on the characteristics of typical 2D photonic crystal microcavity. *Appl. Phys. B* **2007**, *88*, 231–236.

29. Vahala, K. *Optical Microcavities*; World Scientific: Hackensack, NJ, USA, 2004.

30. Englund, D.; Fushman, I.; Vučković, J. General recipe for designing photonic crystal cavities. *Opt. Express* **2005**, *13*, 5961–5975.

31. Sapienza, L.; Thyrrestrup, H.; Stobbe, S.; Garcia, P.D.; Smolka, S.; Lodahl, P. Cavity quantum electrodynamics with Anderson-localized modes. *Science* 2010, *327*, 1352–1355.

4

RESONANT PHOTONIC STATES IN COUPLED HETEROSTRUCTURE PHOTONIC CRYSTAL WAVEGUIDES

J. D. Cox, J. Sabarinathan and M. R. Singh

ABSTRACT

In this paper, we study the photonic resonance states and transmission spectra of coupled waveguides made from heterostructure photonic crystals. We consider photonic crystal waveguides made from three photonic crystals A, B and C, where the waveguide heterostructure is denoted as B/A/C/A/B. Due to the band structure engineering, light is confined within crystal A, which thus act as waveguides. Here, photonic crystal C is taken as a nonlinear photonic crystal, which has a band gap that may be modified by applying a pump laser. We have found that the number of bound states within the waveguides depends on the width and well depth of photonic crystal A. It has also been found that when both waveguides are far away from each other, the energies of bound photons in each of the waveguides are degenerate. However, when they are brought close to each other, the degeneracy of the bound states is removed due to the coupling between them, which causes these states to split into pairs. We have also investigated the effect of the pump field on photonic crystal C. We have shown that by applying a pump field, the system may be switched between a double waveguide to a single waveguide, which effectively turns on or off the coupling between degenerate states. This reveals interesting results that can be applied to develop new types of nanophotonic devices such as nano-switches and nano-transistors.

KEYWORDS

Photonic crystal heterostructures, Nonlinear photonic crystals, Photonic crystal waveguides, Nanophotonics, Resonant tunneling, Coupled waveguides

1. INTRODUCTION

There has been a great deal of interest in photonic crystals [1, 2] due to their ability to manipulate the propagation of light in the same way that semiconductor materials are able to control the propagation of electrons. Photonic crystals are materials with a dielectric constant that varies periodically in one, two or three spatial dimensions, due to which a photonic band gap forms in the structure's photonic dispersion relation. Significant effort has been devoted to the development of new photonic devices made from photonic crystals. Two promising classes of these materials are photonic crystal heterostructures [3] and nonlinear photonic crystals [4–14].

Photonic crystal heterostructures are formed by joining two or more photonic crystals into a single structure, which gives the heterostructure a more complex band structure than that of a raw photonic crystal [3]. Photonic crystal heterostructures have been successfully used to develop devices such as high-quality resonant cavities [15], low-loss waveguides [16] and high-efficiency add-drop filters [17]. Nonlinear photonic crystals possess a PBG that can be shifted, which makes these materials ideal for developing optical switching devices. Recently, optical switching mechanisms due to nonlinear Kerr effects in photonic crystals have been studied both theoretically [4–9] and experimentally [10–14].

Here, we study the photonic resonance states and transmission spectra in coupled waveguides made from heterostructure photonic crystals. Waveguides made by embedding a high refractive index material into a low index material have been widely studied [18]. Some work has also been done on photonic crystal waveguides formed by doping a dielectric material into a photonic crystal. For example, McGurn [19] and Sinha [20] have studied co-directional coupling between two photonic crystal waveguides. Faraon et al. [21] have studied theoretically and experimentally the coupling between photonic crystal waveguides.

In this paper, we consider two linear photonic crystals A and B and a nonlinear photonic crystal C and arrange them as B/A/C/A/B to form the waveguide heterostructure. Crystal parameters for A, B and C are chosen so that the upper band edge of crystal A lies within the band gaps of crystals B and C. Due to this band structure engineering, light is confined within crystal A, which thus act as waveguides. In the confinement direction of the waveguide, photons will occupy discrete quantized states within photonic crystal A, which appear as photonic quantum wells (PQWs). This so-called photonic confinement effect has been demonstrated both theoretically [22–24] and experimentally [25, 26] for various types of PQW systems, analogous to the electronic confinement effect that occurs in semiconductor quantum wells. It has been shown that the phenomenon of resonant tunneling will occur for a PQW with sufficiently thin photonic barriers, whereby an incident photon possessing an energy matching a resonant state of the PQW will undergo perfect transmission through both barriers [22, 23]. Because of this, resonant states appear as sharp peaks approaching unity in the transmission spectrum of a PQW [22–26].

Using the transfer matrix method [22, 24, 27], we have determined the energies of bound photons within the coupled waveguides and calculated the transmission coefficient of the system in the direction of photon confinement. We have found that the number of bound states within the waveguides depends on the width and the well depth of the photonic wells. It has also been found that when both waveguides are far away from each other (or equivalently, the width of photonic crystal C is large), the energies of bound photons within the waveguides are degenerate. However, when the waveguides are brought close to each other, the degeneracy of the bound states is removed due to the coupling between them, which causes these states to split into symmetric and anti-symmetric pairs. We have also investigated the effect of the nonlinearity of photonic crystal C. By applying a pump laser, the photonic band gap of crystal C changes due to the Kerr effect [8]. This change in band structure effectively modifies the energy barrier separating the waveguides, which in turn changes the splitting energy of the coupler. In other words, the energies of the couplers can be switched on and off by applying the pump laser. This reveals interesting results that can be applied to develop new types of nanophotonic devices such as nano-switches and nano-transistors [4–14].

2. THEORY

We consider three photonic crystals A, B and C, where each is composed of dielectric spheres embedded in a dielectric background material. Dispersion relations for these crystals were obtained using the isotropic band structure model proposed by John and Wang in [28], which has been widely used in the literature of photonic crystals to study their optical properties [8, 28,29]. For a photonic crystal described by this model, it is considered that all electromagnetic waves propagating within the crystal encounter the same periodically varying dielectric constant, regardless of their polarization or propagation direction. Due to this simplifying assumption, light propagation within this photonic crystal can be described by a scalar wave equation (1-D Maxwell wave equation) for the electric field. By solving the scalar wave equation, we obtain the dispersion relation for photonic crystal i(i.e. i = A, B or C) as

$$\cos(k_i L_i) = \frac{(n_{s,i}+n_{b,i})^2}{4n_{s,i}n_{b,i}}\cos\left[\frac{\varepsilon_k(2n_{s,i}a_i+n_{b,i}b_i)}{\hbar c}\right] - \frac{(n_{s,i}-n_{b,i})^2}{4n_{s,i}n_{b,i}}\cos\left[\frac{\varepsilon_k(2n_{s,i}a_i-n_{b,i}b_i)}{\hbar c}\right] \quad (1)$$

In Eq. (1),k_i is the Bloch wave vector, L_i is the lattice constant, $n_{s,i}$ and $n_{b,i}$ are the indices of refraction of the spheres and background materials, respectively, a_i and b_i are the radius of the spheres and the distance between the spheres, respectively, and ε_k is the photon energy. Although the isotropic band structure model employed here is an idealization, it leads to qualitatively correct physics and has been shown to exhibit many features of observed and simulated band structures for 3-D photonic crystals [28, 29].

The double waveguide heterostructure is formed by arranging photonic crystals A, B and C in the sequence B/A/C/A/B. We consider that the waveguide is symmetric in the direction of photon confinement, such that opposite layers of photonic wells and barriers have equal thicknesses that are given as d_A, d_B and d_C. It is considered here that the spheres in photonic crystal C consist of a nonlinear dielectric material, such that their refractive index in the presence of an intense pump field is given as [8,30]

$$n_{s,C} = n_0 + n_{nL} I \quad (2)$$

where n_{nL} is a function of the third-order susceptibility, which can be positive or negative depending on the dielectric material being used, and I is the intensity of the pump field laser. When the pump laser is applied, the band gap of photonic crystal C will be shifted due to the Kerr nonlinearity (Fig. 1).

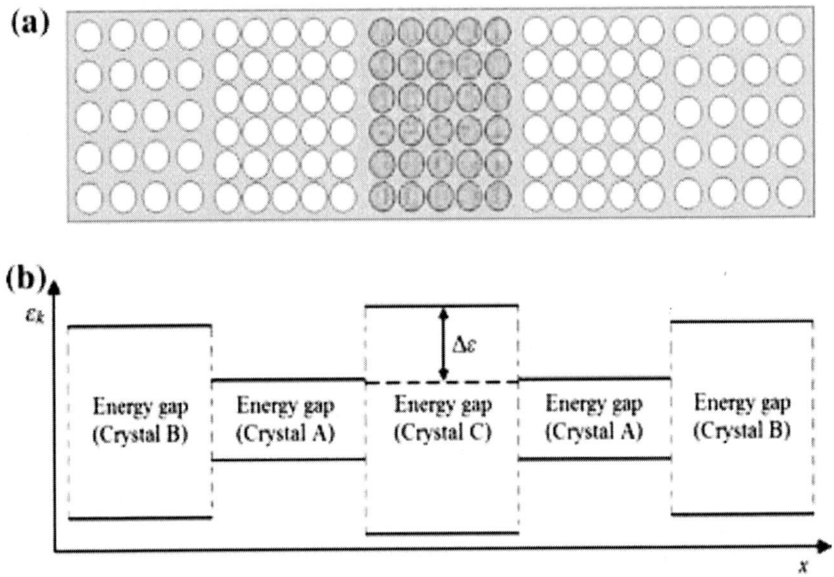

Figure 1 Schematic diagram of the coupled heterostructure photonic crystal waveguide. **a** Cross-sectional view of the waveguide, depicting the photonic crystal arrangement within. The *circles* represent the spheres imbedded in each crystal, which are filled with air in photonic crystals A and B and a nonlinear dielectric material in photonic crystal C. The backbone material of each crystal consists of the same linear dielectric material. **b** Energy band structure of the waveguide in the direction of photon confinement. The *dashed line* located in the central photonic barrier illustrates how the position of the upper band edge of photonic crystal C shifts when the pump field is applied, where Δ ε denotes the change in energy. Note that the lower band edge of crystal C also shifts in a similar manner upon the application of the pump field; this is not depicted here, since we are only concerned with the PQWs formed by the upper band edges of photonic crystals A, B and C

Here, we choose crystal parameters such that the upper band edges of photonic crystals B and C are almost equal, and the upper band edge of photonic crystal A is below that of B and C, but not so far as to form a deep photonic well. This is done to ensure that the photonic wells formed within crystal A are shallow enough so that the shift in the band gap of crystal C will be great enough to eliminate the central photonic barrier. Thus, we can switch the system between a double waveguide and single, wider waveguide by applying a pump field.

It is considered that light in the form of transverse electric waves is incident upon the waveguide heterostructure in the direction of photon confinement. The transmission coefficient of the heterostructure is calculated for the confinement direction using the transfer matrix method [27], which matches the electric field and its first derivative at the interfaces between adjacent photonic crystal layers in order to relate the incident and reflected amplitudes in each layer via transfer matrix equation. The transfer matrix method is ideal for our simulations, because we are considering isotropic photonic crystals, and we are only concerned with light propagation along the direction of photon confinement. In principle, a photonic waveguide does not allow transmission through the outer photonic barriers, but here we allow for transmission to occur by choosing sufficiently thin outer photonic barriers that permit photonic tunneling to occur. We do this in order to study the resonant tunneling effect that occurs as a result, whereby the behavior of the bound photonic states in the waveguide can be studied from its transmission spectra. The energies of the bound states are not affected by the thickness of the outer photonic barriers, so this has no effect on energies of the bound states within the waveguides or their coupling strength. Using the dispersion relation given in Eq. (1), the transmission coefficient was obtained as a function of the incident photon energy and was plotted over the range of energies between the upper band edges of photonic crystals A and B in order to find the resonant states of the system. The transmission spectra were studied for instances when the pump field is on or off in order to observe the switching effect produced.

3. RESULTS

We consider a coupled heterostructure photonic crystal waveguide consisting of crystals A, B and C arranged as B/A/C/A/B. For these crystals, their specific 3-D lattice structures are not be specified, as the isotropic band structure model employed here is independent of the type of lattice considered. Photonic crystals A, B and C each have a dielectric background material consisting of titania (TiO_2), such that $n_{s,A} = n_{s,B} = n_{s,C} = 2.5$ [30, 31]. The spheres in crystals A and B are taken to be filled with air $n_{b,A} = n_{b,B} = 1.00$, whereas the dielectric spheres in crystal C are made of polystyrene, with $n_o = 1.59$ and $n_{nL} = 1.14 \times 10^{-12}$ cm²/W [9]. We choose polystyrene as our nonlinear dielectric material due to its strong and fast Kerr nonlinear optical response [9, 14]. Since the third-order nonlinear susceptibility of titania is on the order of 10^{-15} cm²/W [30], photonic crystals A and B are considered to be linear in our simulations. The radii of the spheres in all three photonic crystals were taken as $a_A = a_B = a_C =$ 125 nm, and the lattice constants for each crystal were taken as $L_A =$ 520 nm, $L_B =$ 510 nm and $L_C =$ 435 nm. The upper photonic band edges of photonic crystals A and B were calculated from Eq. (1) as 0.79273 eV and 0.81830 eV, respectively. When the pump field is off, the upper photonic band edge of photonic crystal C is found to be 0.82349 eV. Transmission spectra of the heterostructure are plotted between the upper band edges of photonic crystals A and B in order to study the resonant photonic states in the waveguide heterostructure.

In Fig. 2, we study the effect of the coupling between the waveguides by varying the thickness of the central photonic barrier (crystal C) while leaving the pump field off. Here, we observe that when the waveguides are near one another, there is a strong spectral splitting effect, and resonant peaks occur in pairs. However, as the waveguides are brought further away from one another, we find that the spectral splitting is diminished, eventually to the point where it can no longer be resolved. The spectral splitting effect observed here occurs due to a coupling between degenerate states of the two waveguides. Because the separation between waveguides is finite, there is some limited interaction between the electric fields of degenerate states. This interaction causes the degenerate states of the system to split into symmetric and anti-symmetric pairs, analogous to molecular bonding and anti-bonding pairs [26]. Thus in Fig. 2, each pair of resonant peaks corresponds to a degenerate state shared by the two waveguides, and for each pair, the lower-energy peak corresponds to the symmetric state, while the higher-energy peak is the anti-symmetric state. For degenerate states with higher energy, we observe that a proportionately stronger degree of spectral splitting occurs, which is due to enhanced photonic barrier penetration (and thus stronger coupling) at these energies. Thus, by changing the waveguide separation, the strength of the coupling between degenerate states can be varied, and the system can be switched between single and double bound states. From Fig. 2, we also note that for all resonant peaks, there is an associated degree of spectral broadening that increases with photon energy. This is attributed an enhanced tunneling rate for photons with energies nearer to the upper band edges of the photonic barriers [23].

The Kerr nonlinearity of photonic crystal C is investigated in Fig. 3, where numerical simulations of the energy band structure have been performed while varying the intensity of an applied external pump field. Here, we only consider a range of pump field intensities up to 10^{11}W/cm^2, which effectively adds a small contribution to the index of refraction for the polystyrene spheres in photonic crystal C, as described by Eq. (2). Note that these pump fields are sufficiently small so that their effect on the index of refraction of titania is negligible, and so photonic crystals A and B are linear. In Fig. 3, the energies of the upper photonic band edges of photonic crystals A, B and C are shown. Bound photonic states within the waveguide heterostructure have energies between the upper band edges of crystals A and B. From Fig. 3, we see that by applying a sufficiently intense pump field to the structure, the upper band edge of photonic crystal C can be shifted to a level slightly below that of crystal A, which effectively removes the central photonic barrier separating the two waveguides. Essentially, this means that the system can be switched between a double waveguide and a single, wider waveguide by applying the pump field.

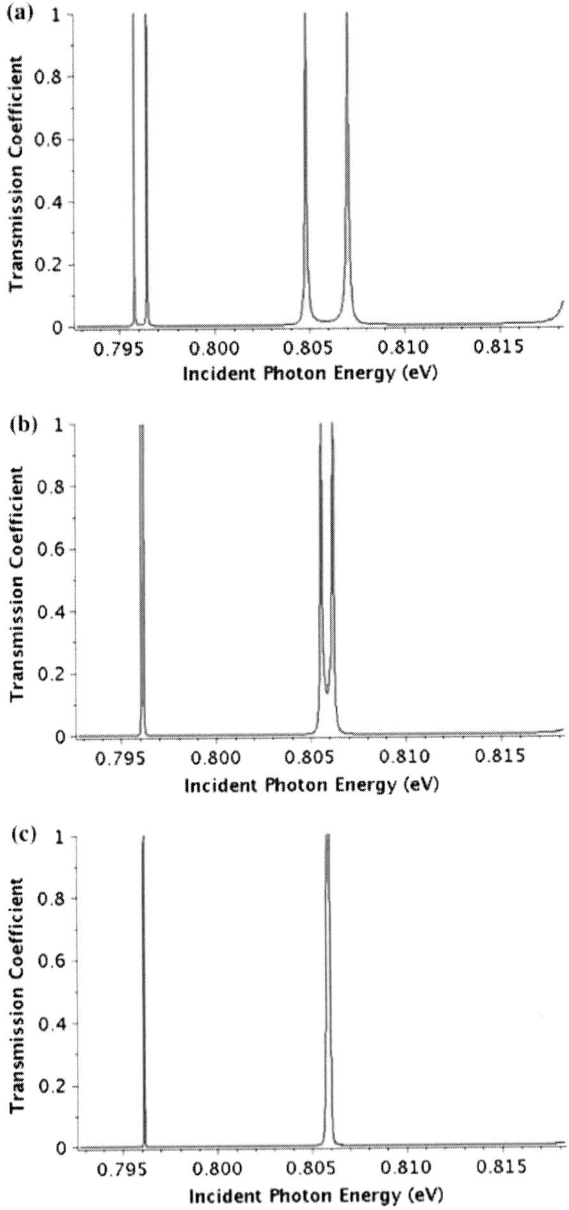

Figure 2 Transmission spectra for the heterostructure waveguide in the direction of photon confinement, where the central barrier width

is varied to observe its effect on the coupling between the waveguides. For all three plots, the outer photonic barriers were taken as $d_B = 5L_B$ and the widths of the wells were $d_A = 15L_A$. Here, the effect of the pump field was not considered. The barrier widths were varied as $d_C = 5L_C$ (**a**),$d_C = 10L_C$ (**b**) and $d_C = 15L_C$ (**c**). When the waveguides are close together, strong coupling of degenerate states enhances the energy-splitting effect. As the waveguides are separated, the interaction between degenerate states diminishes, and the energy splitting cannot be resolved

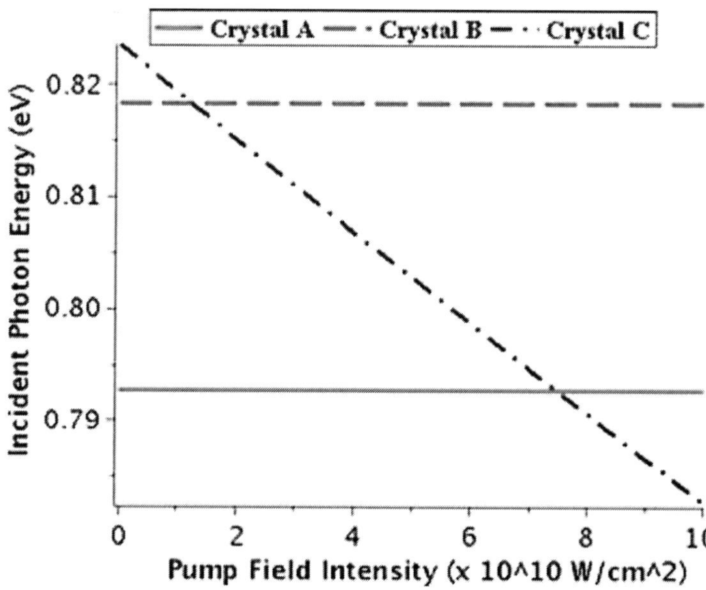

Figure 3 Energy of the upper photonic band edge of photonic crystals A, B and C as the intensity of the pump field is varied. Since crystals A and B are essentially linear for this range of pump field intensity, their upper band edges are unaffected by the pump field. We see here that by applying a pump field with intensity around 8 × 10^{10}W/cm^2, the upper band edge of crystal C can be shifted to just slightly below that of crystal A, effectively removing the photonic barrier separating the two waveguides in the heterostructure

In Fig. 4, we consider transmission spectra of the coupled waveguide heterostructure for cases where the external pump field is on or off (i.e. applied or not). Here, we have considered a pump field with an intensity of 8×10^{10} W/cm^2, which lowers the energy of the upper band edge of photonic crystal C to 0.79064 eV. This is only slightly below the upper band edge of crystal A (0.79273 eV) and causes photonic crystal C to function as a photonic well rather than a barrier. In Fig. 4, we see that the transmission spectrum has changed dramatically by applying the pump field, where there is a greater number of resonant peaks that no longer occur in split pairs. By applying the pump field, the system has switched from a coupled waveguide structure to that of a single, wider waveguide consisting of the two photonic crystals A and the central layer of photonic crystal C. Since the number of bound photonic states depends on the width of the waveguide [23], we see a greater number of resonant peaks in the transmission spectrum. Thus, we have shown here that by applying a pump field, the system can be switched between single and double waveguide states (Fig. 4).

4. CONCLUSIONS

Here, we have considered coupled heterostructure photonic crystal waveguides made from two photonic crystals A and B and another nonlinear photonic crystal C, where the heterostructure is denoted as B/A/C/A/B. Due to the band structure engineering, light is confined within photonic crystal A by crystals B and C, causing two waveguides to form in crystal A. Using the transfer matrix method, we have studied the resonant photonic states within the waveguides by calculating the transmission coefficient of the system in the direction of photon confinement. We have found that the number of bound photonic states within the waveguides can be controlled by varying the width and/or well depth of photonic crystal A. It was also found that when the waveguides are far away from each other, the energies of the bound photons are degenerate. However, by bringing the waveguides near one another, the degeneracy of the bound states is removed due to coupling between them, causing these states to undergo a twofold energy-splitting effect. Here, we have also investigated the nonlinearity of photonic crystal C. By applying a pump laser, the photonic band gap of crystal C changes due to the Kerr nonlinearity. The change in band structure causes the central photonic barrier to disappear, switching the system between single and double waveguide states. Essentially, this means that the energy-splitting effect can be turned off by applying a pump laser. It is expected that these results can be applied to develop new types of nanophotonic switching devices.

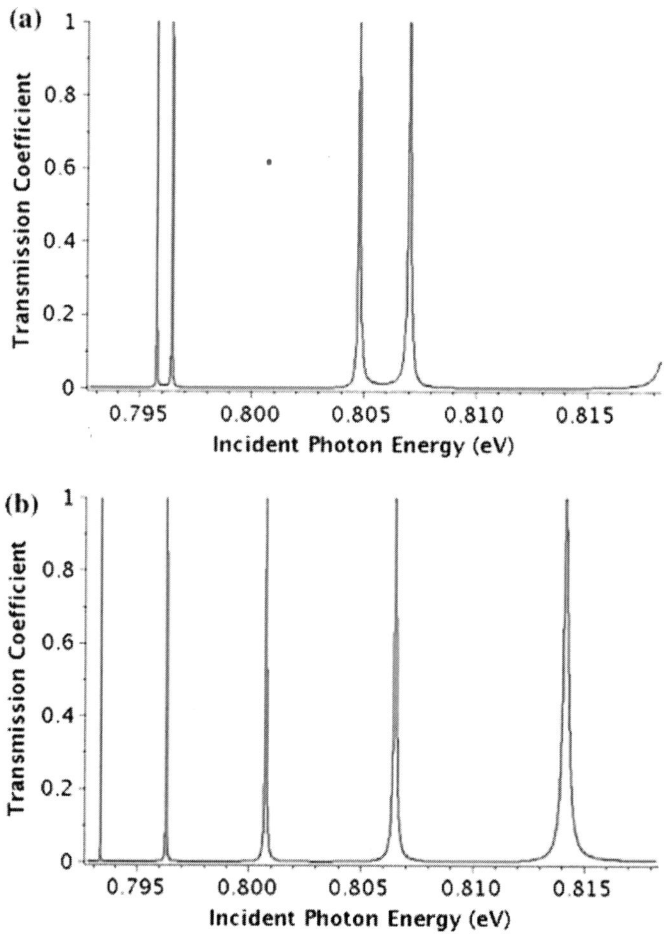

Figure 4 Transmission spectra for the heterostructure waveguide in the direction of photon confinement, where the pump field is off (**a**) and on (**b**). For these plots, the outer photonic barriers were taken as $d_B = 5L_B$, the widths of the wells were $d_A = 15L_A$, and the central (nonlinear) barrier was taken as $d_C = 5L_C$. By applying the pump field, the system is switched between a double waveguide and a single, wider waveguide. This effectively switches the energy-splitting effect "off" or "on"

REFERENCES

1. E. Yablonovitch, Phys. Rev. Lett. 58, 2059 (1987)

2. S. John, Phys. Rev. Lett. 58, 2486 (1987)

3. E. Istrate, E.H. Sargent, Rev. Mod. Phys. 78, 455 (2006)

4. M. Scalora, J.P. Dowling, C.M. Bowden, M.J. Bloemer, Phys. Rev. Lett. 73, 1368 (1994)

5. P.R. Villeneuve, D.S. Abrams, S. Fan, J.D. Joannopoulos, Opt. Lett. 24, 2017 (1996)

6. P. Tran, J. Opt. Soc. Am. B 14, 2589 (1997)

7. M. Soljacˇic´, C. Luo, J.D. Joannopoulos, Opt. Lett. 28, 637 (2003)

8. M.R. Singh, R. Lipson, J. Phys. B: At. Mol. Opt. Phys. 41, 015401 (2008)

9. Y. Lin, F. Qin, F. Zhou, Z.Y. Li, J. Appl. Phys. 106, 083102 (2009)

10. A. Hache´, M. Bourgeois, Appl. Phys. Lett. 77, 4089 (2000)

11. S.W. Leonard, H.M. van Driel, J. Schilling, R.B. Wehrspohn, Phys. Rev. B 66, 161102(R) (2002)

12. M. Shimizu, T. Ishihara, Appl. Phys. Lett. 80, 2836 (2002)

13. D.A. Mazurenko et al., Phys. Rev. Lett. 91, 213903 (2003)

14. Y. Liu et al., Appl. Phys. Lett. 95, 131116 (2009)

15. B.S. Song, S. Noda, T. Asano, Y. Akahane, Nat. Mater. 4, 207 (2005)

16. S.Y. Lin, E. Chow, S.G. Johnson, J.D. Joannopoulos, Opt. Lett. 25, 1297 (2000)

17. H. Takano, B.S. Song, T. Asano, S. Noda, Opt. Express 14, 3491 (2006)

18. Y. Nakayama et al., Nature 447, 1096 (2007)

19. A.R. McGurn, Phys. Rev. B 65, 075406 (2002)

20. Y. Nagpal, R.K. Sinha, Microw. Opt. Technol. Lett. 43, 47–50 (2004)

21. A. Faraon et al., Appl. Phys. Lett. 90, 073102 (2007)

22. J. Zi, J. Wan, C. Zhang, Appl. Phys. Lett. 73, 2084 (1998)

23. Y. Jiang, C. Niu, D.L. Lin, Phys. Rev. B. 59, 9981 (1999)

24. C.S. Feng et al., Solid State Comm. 135, 330 (2005)

25. S. Yano et al., Phys. Rev. B 63, 153316 (2001)

26. S.H. Xu et al., Photonics and Nanostructures Fundam. Appl. 4, 17–22 (2006)

27. A. Yariv, P. Yeh, Photonics: optical electronics in modern communications (Oxford University Press, Oxford, 2007)

28. S. John, J. Wang, Phys. Rev. B 43, 12772 (1991)

29. P. Lambropoulos, G.M. Nikolopoulos, T.R. Nielsen, S. Bay, Rep. Prog. Phys. 63, 455 (2000)

30. R.W. Boyd, Nonlinear Optics, 3rd edn. (Academic Press, New York, 2008)

31. J. Wijnhoven, W.L. Vos, Science 281, 802 (1998)

5

LIQUID CORE ARROW WAVEGUIDES: A PROMISING PHOTONIC STRUCTURE FOR INTEGRATED OPTOFLUIDIC MICROSENSORS

Genni Testa *, Gianluca Persichetti and Romeo Bernini

Istituto per il Rilevamento Elettromagnetico dell'Ambiente, Consiglio Nazionale delle Ricerche (IREA-CNR), Via Diocleziano 328, 80124 Naples, Italy

ABSTRACT

In this paper, we introduce a liquid core antiresonant reflecting optical waveguide (ARROW) as a novel optofluidic device that can be used to create innovative and highly functional microsensors. Liquid core ARROWs, with their dual ability to guide the light and the fluids in the same microchannel, have shown great potential as an optofluidic tool for quantitative spectroscopic analysis. ARROWs feature a planar architecture and, hence, are particularly attractive for chip scale integrated system. Step by step, several improvements have been made in recent years towards the implementation of these waveguides in a complete on-chip system for highly-sensitive detection down to the single molecule level. We review applications of liquid ARROWs for fluids sensing and discuss recent results and trends in the developments and applications of liquid ARROW in biomedical and biochemical research. The results outlined show that the strong light matter interaction occurring in the optofluidic channel of an ARROW and the versatility offered by the fabrication methods makes these waveguides a very promising building block for optofluidic sensor development.

KEYWORDS

liquid ARROW; optofluidics; microfluidics; optical sensors; integrated silicon technology

1. INTRODUCTION

Optofluidics is an emerging research field that combines the advantages of microfluidics and optics on the same platform towards highly functional and compact devices [1,2]. More precisely, optofluidic approaches provide high-yield integration of fluidics by including fluidic elements as parts of the photonic structure. This solution confers to the optofluidic systems a high level of compactness and reconfigurability as the fluids provide an effective means to tune and reconfigure the photonic architecture.

In the field of optical sensor technologies, optofluidics offers innovative design solutions in order to integrate all the required microfluidic and optical functionalities on a single chip for high-throughput analysis. Optofluidic sensors have been extensively developed for biomedical and biochemical analysis for healthcare purposes (*i.e.*, diagnostic), pharmaceutical research, and environmental monitoring [3]. Whatever optical method is used, highly-sensitive detection requires a very careful design of both optical and microfluidic parts in order to push the device sensitivity to the limit of the theoretical capability. Furthermore, innovative design strategies need to be developed to overcome the reduced interaction length owing to devices size miniaturization. In this context, optofluidics answer to the growing need for miniaturized and portable sensing devices, highly integrated, and extremely sensitive [4].

Optofluidic waveguides represent one of the simplest examples of optofluidic integration, as the optical and fluidic structures are tightly integrated into each other. In particular, an optofluidic waveguide is a photonic structure able to perform optical confinement of the light through a fluid [5]. Among others, liquid cores antiresonant reflecting optical waveguides (ARROWs) have been revealed as a very powerful building-block for optofluidic sensors [6]. In an ARROW the light is guided through the liquid core by means of highly reflective mirrors suitably realized on the core sidewalls. Since both the light and the liquid are guided through the same microchannel, ultra-high sensitivity can be achieved by exploiting the direct interaction of the entire optical waveguide mode with the fluid samples. Such a strong optical coupling, combined with the ultra-small liquid volume, enables sensitivity down to single molecule level, which is the ultimate goal of analytical methods. ARROWs can be easily fabricated using silicon technologies; moreover, like others optofluidic waveguides, such as the photonic crystal [7,8] and the slot waveguides [9,10], ARROWs are, in essence, planar and, hence, highly attractive for planar optofluidic integration [11,12]. As a matter of fact, starting from the first demonstration of the ARROWs' potentialities in the field of optical sensing, there has been a growing interest for them as basic elements in optofluidic chip for sensing applications.

In the following, we report on optofluidic sensors realized with liquid ARROWs. Starting from the description of the working principle and fabrication methods of liquid ARROWs, we review the state of the art of on chip integration for sensing applications, highlighting specific advantages gained through the optofluidic integration.

2. OPERATION PRINCIPLES AND SIMULATIONS

2.1. 1-D Liquid ARROW

ARROW waveguides enable light confinement and propagation thanks to interference phenomena occurring in the cladding structure, made up of alternating high- and low- refractive index (RI) thin layers. The working principle of an ARROW waveguide can be easier understood by considering a schematic one-dimensional structure, as shown in Figure 1. For simplicity, we consider only two cladding layers with refractive index n_1 and n_2 ($n_1 > n_2$). Light propagates through the core by means of Fresnel reflections at the cladding interfaces. The entrapped light in each cladding layer experiences interference due to repeated reflections at cladding interfaces. In order to intensify the light reflected back into the core, each interference cladding must be designed to fulfill a specific condition. By considering the analogy of a cladding layer with a Fabry–Pérot cavity, such condition corresponds to the anti-resonance state of the cavity, which performs the maximum reflectivity. The reflectivity of the multilayer stack constituted by multiple interference claddings, generally called Fabry–Pérot mirror, can be very high, up to 99%.

A detailed description of the ARROW confinement phenomenon is reported in [13], where the phase accumulated by the fundamental mode at each round trip in the interference claddings is clearly defined. By setting the specific condition of antiresonance, a very useful expression to calculate the optimal cladding thicknesses d_1 and d_2, can be derived:

$$d_{1,2} = \frac{\lambda}{4}\left[n_{1,2}^2 - n_{eff}^2\right]^{\frac{-1}{2}}(2N+1) \qquad N = 0, 1, 2, \ldots \tag{1}$$

where λ is the working wavelength, $n_{1,2}$ is the refractive index of the cladding layers and n_{eff} is the effective index of the propagation mode. For the fundamental mode, n_{eff} can be approximately given by:

$$n_{eff} = \frac{\beta}{\beta_0} \simeq n_c \left[1 - \left(\frac{\lambda}{2n_c d_c}\right)^2\right]^{\frac{1}{2}} \quad (2)$$

where $\beta_0 = 2\pi/\lambda$, β is the propagation constant, n_c and d_c are the refractive index and the thickness of the core, respectively. If $\lambda/d_c \ll 1$ (e.g., large core waveguides), Equation (2) can be simplified to $n_{eff} \approx n_c$. As long as the operating wavelength and the core and cladding materials are conveniently chosen, reduced propagation losses can be obtained by using Equations (1) and (2) to calculate the cladding thickness.

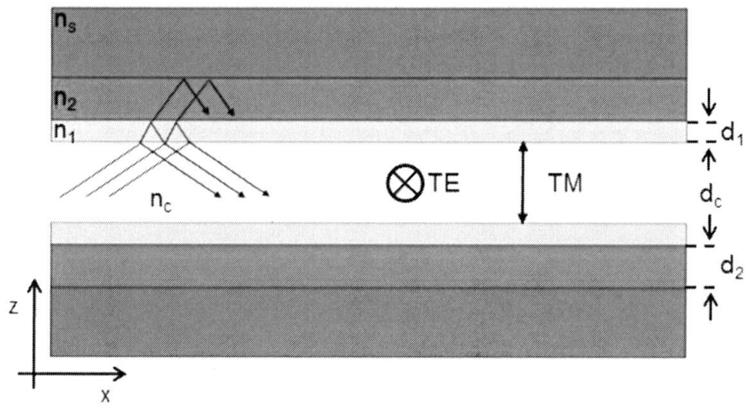

Figure 1. 1-D structure of ARROW waveguide.

It should be noted that, unlike photonic crystal waveguides, in ARROW waveguides the high- and low-index alternating layers do not require a high periodicity in the structure [14]. This characteristic simplifies the waveguides manufacturing and strongly relaxes the fabrication constrains.

ARROWs are leaky waveguides and require a very careful design to operate in a single-mode condition, which is a fundamental requirement for sensing applications, *i.e.*, for the fabrication of interferometric devices. Single-mode operation can be accomplished by exploiting the weaker reflectivity experienced by higher order modes upon reflections at cladding interfaces. Low-order modes propagate with glancing angles at cladding interfaces, hence the Fresnel coefficient is quite high and the loss is low, whereas higher-order modes have larger incident angles and hence experience higher losses. An extensive investigation of the ARROW attenuation losses has been reported in [15]. The authors provide very useful analytical expressions that, with a good approximation, describe the mode losses for both TE and TM-polarized modes. A simple calculation form analytical expression given in [15] permits to calculate the ARROW attenuation coefficient for the TE and TM fundamental modes in the following, very useful, approximated form:

$$\begin{cases} \alpha_{TE} = \dfrac{0.5\lambda^4\left[1+\dfrac{4d_c^2}{\lambda^2}\left(n_2^2-n_c^2\right)\right]}{n_c d_c^5\left(n_1^2-n_c^2\right)\sqrt{\left(n_s^2-n_c^2\right)}} \\ \alpha_{TM} = \left(\dfrac{n_1^2 n_s}{n_c n_2^2}\right)^2 \alpha_{TE} \end{cases} \tag{3}$$

which is valid when $d_c >> \lambda$; n_s is the refractive index of the substrate. By assuming Si_3N_4 (n_1 = 2.01) and SiO_2 (n_2 = 1.44) for the first and second layer, Si (n_s = 3.8) for the substrate, and water core (n_c = 1.33), the proportional coefficient between α_{TM} and α_{TE} in Equation (3) is as high as 30. Hence the propagating modes experience attenuation coefficients that strongly depend upon their state of polarization. In particular, propagation of TE-polarized light is always privileged due to the lower attenuation coefficients. Additionally, in this case this behavior originates from the light confinement mechanism, based on Fresnel reflections at core and cladding boundaries. Fresnel reflection depends on the polarization of the incident light and TM reflection coefficient is always lower with respect to TE reflection.

Moreover, as a general rule, from Equation (3) the attenuation coefficients decrease by increasing the core dimensions. This implies that the core size must be suitably chosen in order to reduce fundamental mode loss but also to minimize the contribution of higher order modes to the total transmitted power. The degree of suppression of these modes improves as the core width decreases [16]. A further very useful general rule can be retrieved: in order to reduce the attenuation coefficients, the refractive index of the first cladding (n_1) should be higher than the core index, while the refractive index of the second cladding (n_2) as closer as possible to the core index (n_c). These requirements impose very important limitations to the choice of the cladding materials. For this reason typical materials are: silicon nitride (n_1 = 2.01–2.2), titanium dioxide (n_1= 2.49), and tantalum dioxide (n_1 = 2.3) for the first high index cladding and silicon dioxide (n_2 = 1.46) for low index cladding, as they provide proper index contrast, for water core waveguides, and simple fabrication by means of silicon compatible process.

By combining these intrinsic properties along with an accurate design strategy, ultra-low loss, single mode linearly polarized ARROWs operation can be accomplished and exploited to realize high-performance interferometric devices.

2.2. 2-D Liquid ARROW

In a practical ARROW waveguide, the core must be completely surrounded by the cladding layers to accomplish light-confinement in both lateral and vertical direction. One possible ARROW structure for both vertical and horizontal confinement is illustrated in Figure 2.

A simple estimation of the attenuation coefficients in 2-D ARROW waveguides can be performed by using the one dimensional analytical model [11,15]. In a paraxial approximation the problem of the 2-D ARROW confinement can be separated into two 1-D problem for the lateral and transverse direction, in this way the evaluation of the total losses can be obtained from the addition of the resulting 1-D losses. In Figure 3 the losses for the fundamental (blue line) and first order mode (red line) *versus* wavelength calculated by using approximated analytical model, are shown. Hence, as anticipated, for a fixed core dimension higher order modes attenuate faster than the fundamental mode. From the figure the typical transmitted spectrum of an ARROW waveguide can also be foreseen. It is characterized by a broad spectral band, corresponding to the range of wavelengths which experience destructive interference in the claddings. The wavelength dependence of the propagation losses could be used to tailor the ARROW spectral response in sensing applications, *i.e.*, to filter out unwanted optical wavelengths from reaching the detector.

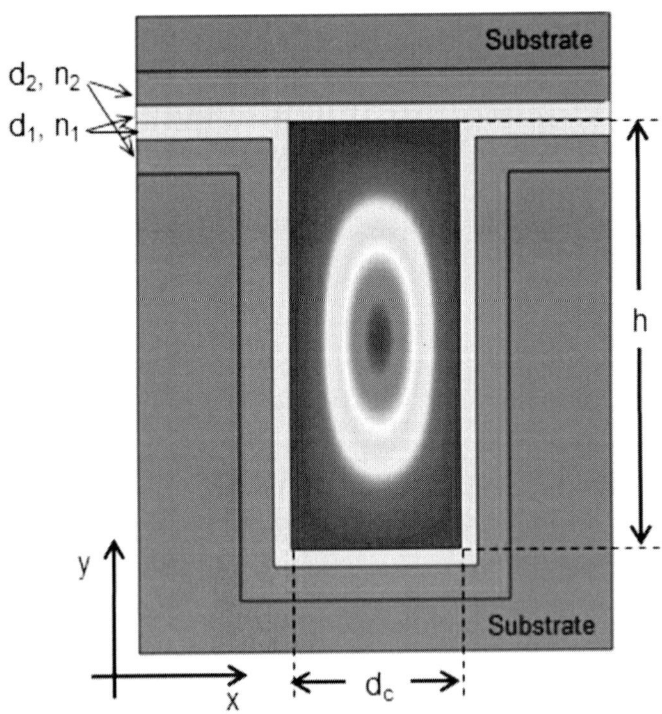

Figure 2. Cross-section of a 2-D liquid ARROW waveguide with simulated power distribution of the fundamental mode.

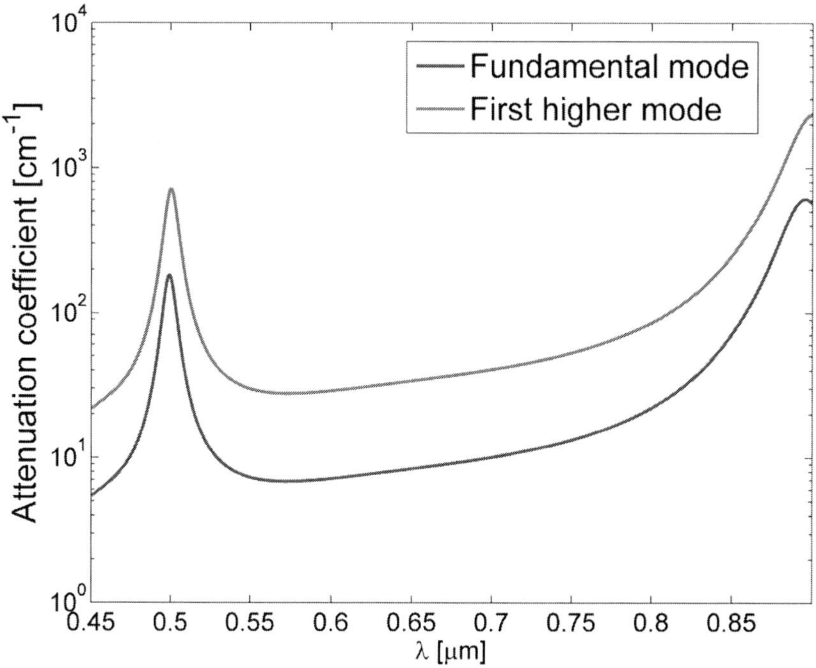

Figure 3. Attenuation coefficient of the fundamental mode (E_y^0) and the first order mode (E_y^{12}) *versus* the wavelength in water-core ARROW waveguide with d_1 = 290 nm,d_2 = 260 nm, d_c = 5 μm and h = 10 μm (Figure 2).

In 2-D ARROW design, the influence of the input polarization on waveguide losses should be taken carefully into account. If incident light is polarized along the *y*-axis (Figure 2) it is reflected upon lateral cladding as TE-polarized light whereas it is TM-polarized with respect to the vertical ARROW confinement. Hence, TE-polarized input light along *x*-axis will also contribute to the total losses as a TM polarization with respect to *y*-axis. By considering the polarization dependence of ARROW losses, these considerations suggest that the polarization of the propagating modes can be selected by suitably designing the core shape. If one of the core dimension is set wider (*i.e.*, a rectangular core shape), modes polarized along the longer side experience lower TM propagation losses with respect to the modes polarized along the crossing direction, *i.e.*, the longer core side identifies the direction of the supported polarization.

More accurate loss evaluation requires 2-D modes numerical analysis. A first detailed numerical investigation of 2-D hollow core ARROWs for sensing application is reported in [6]. The authors simulated ARROWs propagation by using finite element method (FEM) numerical analysis. The propagation characteristics of an ARROW with a hexagonal core cross section have been studied for different diameters of the core, filled with air or liquid substances. The authors also investigated the attenuation losses characteristics of the fundamental mode *versus* the detuning of the thickness of the first and second cladding layer from its antiresonant value. As it was found for solid ARROW [13], their study confirms that the attenuation losses are almost unaffected by the second layer thickness changes in a range of ±30 nm, whereas it is very sensitive to first cladding thickness, increasing very rapidly for thickness detuning of about ±10 nm. These results provide very useful rules for waveguides fabrication. 2-D numerical analysis of ARROWs modes characteristics have been also performed by using finite difference methods (FDM) [16,17].

3. FABRICATION AND CHARACTERIZATION

Currently, ARROWs are fabricated with two main different approaches: by bulk micromachining process or by surface micromachining process. In the first case, the ARROW is constituted by two halves of silicon substrate with the core etched in one of the halve part. The two processed silicon substrates are finally bonded after cladding depositions on both halves [17,18]. In the second case, the core is obtained by depositing and patterning a sacrificial material on a single silicon substrate in between top and bottom claddings. The sacrificial material is removed by etching at the end of process [19,20]. In both processes cladding layers are deposited using deposition techniques like plasma enhanced chemical vapor deposition (PECVD), low pressure chemical vapor deposition (LPCVD) or atomic layer deposition (ALD).

Both fabrication processes have some advantages and disadvantages. Bulk micromachining is a faster process but it needs a non-standard wafer bonding between high index materials like silicon nitride or titanium dioxide. However it enables a flexible design, as the top substrate can be substituted, allowing for the implementation of hybrid configuration, as it will be illustrated in the following. On the other hand, sacrificial etching of long narrow micro channels is very time consuming and it requires controlled manufacturing in order to achieve 100% fabrication yields [20]. Nevertheless, this approach allows a simple monolithic integration with solid core exciting/collecting waveguides. As it will be showed in the following, in both cases ARROW waveguides have been successfully produced and applied to fabricate very high performance and promising optofluidic devices for sensing applications.

Single-mode ARROWs with hollow core have been firstly demonstrated by Yin *et al.* [21]. Hollow core dimensions were 3.5 μm by 12 μm, obtained by sacrificial etching of SU-8 (Figure 4a). For low-loss propagation, four antiresonant claddings were fabricated by alternating silicon oxide and nitride layer depositions on a silicon substrate using PECVD. The waveguide was designed to operate at the wavelength of 785 nm and with $n_c = 1$. The authors demonstrated single-mode operation with propagation loss as low as 6.5 cm^{-1}. The same waveguide was also characterized with a liquid-filled core ($n_c = 1.33$), showing waveguide attenuation coefficient of 2.4 cm^{-1} at $\lambda = 633$ nm [19]. Successive improvements in the fabrication process have been proposed by Yin *et al.* [22] to reduce the waveguide losses as low as 2.6 cm^{-1} for 15×3.5 μm^2 core dimensions ($n_c = 1$). In this work the authors also discussed the effect on the waveguides losses of different growing rate over the horizontal and vertical core walls, which is intrinsic in the PECVD process. This effect causes that the vertical and horizontal layers do not simultaneously fulfill the antiresonance condition, deteriorating the waveguide performances. Even if this effect can be compensated by appropriate layer design [23], it can lead to non-optimal waveguide losses. In the work of Testa *et al.* [24] the ALD process has been proposed to overcome these limitations. ARROW waveguide was fabricated by using bulk micromachining, starting with the realization of the microchannels by etching 5 μm in the silicon wafer. Alternating layer of silicon dioxide and titanium dioxide were deposited on the core sidewalls by using LPCVD and ALD, respectively. One of the advantages of the ALD technique is the possibility to deposit very thin cladding layer (sub-100 nm) with an excellent conformality, reproducibility, and with a high precision over the resulting thickness. The waveguides were designed to operate on single mode at λ = 635 nm with a water filled core (n_c = 1.33). The fundamental mode loss of 5.22 cm^{-1} was obtained with only two antiresonant layers. In Figure 5 the measured and simulated transmitted spectra of a 1.5 cm long ARROW waveguide fabricated by ALD are shown. Simulations have been performed by using by using a commercial software (FIMMWAVE, Version 5.2, © Photon Design) and FDM method. In the simulation model we have taken into account the wavelength dependence of the materials composing the waveguiding structure and the liquid filling the core (methanol). As Figure 5 shows, there is a discrepancy between the simulated and measured spectrum at low wavelength (below 400 nm). This discrepancy is due to the difficulty to obtain an accurate model for the core and the cladding layers refractive indexes (both real and imaginary parts) at wavelengths lower than 400 nm. In particular, as it can be noticed in figure, the spectral region at around 370 nm corresponds to the resonance condition in the cladding layers. As it is demonstrated in the [25], the wavelength position of the minimum in the transmitted spectrum (resonance in the claddings) is greatly influenced by the core and cladding refractive indexes.

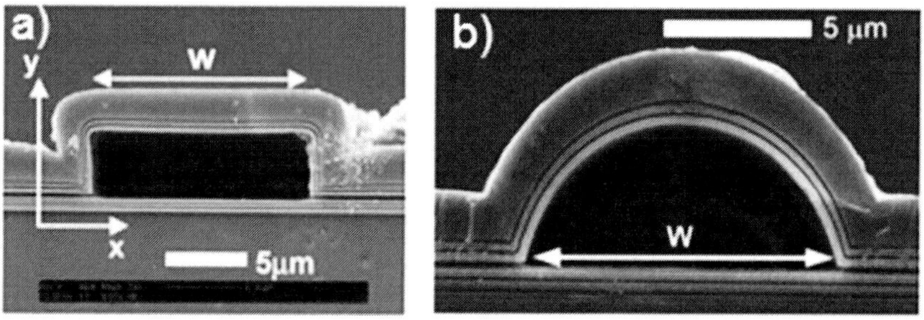

Figure 4. Scanning electron microscope (SEM) images of (**a**) hollow-core ARROWs with rectangular and (**b**) arch-shaped cross sections fabricated by surface micromachining process. Reprinted with permission from [27]. Copyright (2005) OSA.

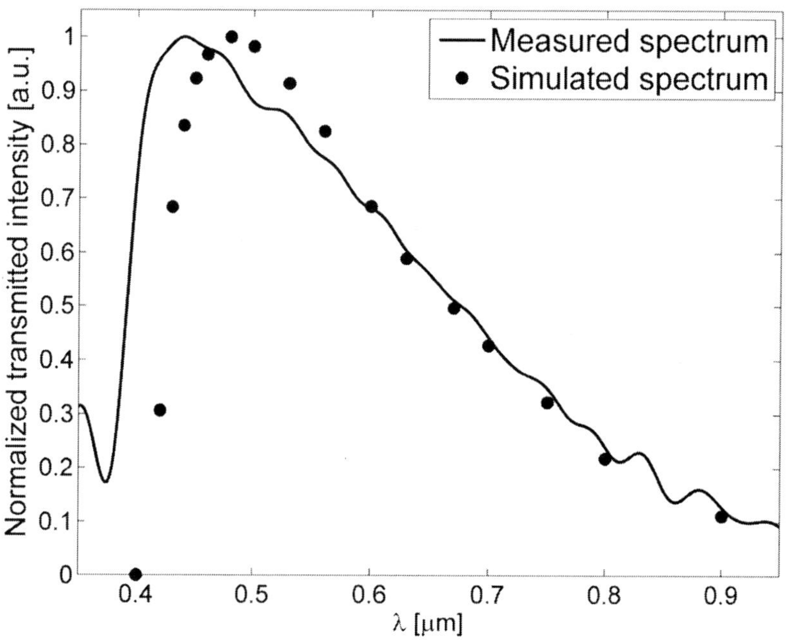

Figure 5. Measured and simulated transmitted spectrum from a single mode ARROW waveguide fabricated by ALD with d_1 = 75.4 nm (TiO_2), d_2 = 262 nm (SiO_2), d_c = 10 μm,h = 5 μm, n_c = 1.32 (methanol-filled core).

Single-mode ARROWs with arc-shaped core were also demonstrated [26] (Figure 4b). The authors fabricated such structures by sacrificial etching and demonstrated their ability to overcome the mentioned limitation of the PECVD process. Further important advantages compared with rectangular ARROWs are the higher mechanical stability and the smoother core walls. Optical characterization of such structures with four antiresonant claddings was also carried out [27]. A low attenuation coefficient of 0.26 cm^{-1} was obtained at $\lambda = 633$ nm with a high refractive index liquid core (ethylene glycol, $n_c = 1.43$).

Single mode hybrid silicon-polymer ARROW (h-ARROW) waveguide has been achieved by Testa *et al.* [28]. The authors exploited the advantage of bulk micromachining to separately process two silicon wafers, which were then bonded at the end of the cladding deposition. H-ARROW was obtained by substituting the upper silicon part with a single polydimethylsiloxane (PDMS) layer. Slightly increased optical losses were expected and measured due to the weaker optical confinement at PDMS-core interface compared with the stronger confinement enabled by antiresonant claddings. However, thanks to a proper design, this effect was minimized and fundamental mode loss of 6.18 cm^{-1} at $\lambda = 633$ nm ($n_c = 1.33$) was obtained. A detailed study of h-ARROW confinement and loss improvement is reported in [16]. This approach circumvents the need of non-standard wafer bonding step between high index materials like silicon nitride or titanium dioxide, which is substituted by a very simple temporary or permanent bonding between silicon nitride and PDMS.

Single-mode liquid ARROWs with high performance have been demonstrated by using the two above-mentioned micromachining processes. Surface micromachining is very promising as it enables monolithic integration with exciting and collecting waveguides. As it can be envisioned, other components can be fabricated and integrated on the same platform in order to add further interesting functionalities for complete lab-on-a-chip realization. Bulk micromachining promises very high flexibility enabled by hybrid approaches. Polymer materials exhibit very favorable properties for both photonic and microfluidic technologies; moreover, thanks to the related inexpensive fabrication procedures, they show a great potential for economic mass production of lab-on-a-chip devices.

4. ARROW-BASED DEVICES AND APPLICATIONS

Optofluidic waveguides have been demonstrated as very promising candidates for high performance optical microsensors [3,5]. Liquid ARROWs play a prominent role in this context, enabling planar integrated fabrication approaches based on the currently available and well-established silicon compatible technology.

Currently, the optical detection methods mainly applied in optical microsensors includes spectroscopic techniques like absorption, fluorescence, and Raman scattering. These approaches have demonstrated to be very effective for highly selective and sensitive detection of molecules down to the single molecule level. Additionally, the detection based on refractive index changes is widely used and have demonstrated a great potential for label-free detection of biomolecules with very high sensitivity.

In the following, we describe optofluidic design approaches for optical microsensors realized with liquid ARROWs.

4.1. Liquid ARROW Waveguides as Optofluidic Sensors

The strong light matter interaction occurring in the liquid core of an ARROW waveguide has been exploited to realize simple but effective and sensitive microfluidic sensors.

In the work of Campopiano *et al.* [25] the first demonstration of an integrated optical bulk refractometer based on multimode liquid ARROW has been proposed. For fluid injection, inlet, and outlet openings were fabricated in the silicon substrate. A multimode ARROW with large core section of 130 × 130 μm² was used. The exciting and collecting optical fibers were inserted directly into the core, resulting in a highly compact and robust self-aligned configuration. Very notably, the waveguide, itself, constitutes the sensor. The sensing principle was based on the intrinsic property of the ARROW: the shift of the minimum of the transmitted spectrum upon core refractive index changes. Good bulk sensitivity of 555 nm/RIU (RIU: refractive index unit) and linear response, with a limit of detection (LOD) of 9×10^{-4} RIU has been demonstrated. The same configuration has been also used to realize an integrated long path absorbance cell for colorimetric detection of specific protein in water solution [17]. The authors used a Bradford assay for colorimetric determination of bovine serum albumin, showing an LOD of about 1 μg/mL with a sample volume of only 0.34 nL and a 15 mm-long-cell. This example was the first experimental demonstration of the sensing capability of a simple straight liquid core ARROW.

Multimode liquid ARROWs have also been employed to realize a flow cytometer for the analysis of fluorescently-labeled human T leukemia cells (Jurkat) [29]. The excitation light for cells interrogation propagated in the optofluidic channel thanks to the ARROW confinement, resulting in a highly compact and efficient optical interrogation scheme. The cells stream was hydrodynamically focused in the center of the channel where two optical fibers were arranged orthogonally for collecting the emitted fluorescent signals.

However, multimode ARROW with large core section area is not preferred to reach ultra-high sensitivity due to the large sample volume that typically causes strong background signal. This objective has been met by using single mode liquid ARROW. Thanks to the core section of a few microns, they provide a very small excitation volume, which is attractive for detection down to single molecule sensitivity.

In the work of Yin *et al.* [23], a liquid ARROW with picoliter core volume has been optimized and applied for highly efficient fluorescence detection of dye molecules, demonstrating an LOD corresponding to 490 molecules in the core. The potential for single molecule sensitivity by surface-enhanced Raman scattering (SERS) detection has been also demonstrated by Measor *et al.* [30] using a simple straight single mode ARROW.

The sensors above described can be considered as the first generation of ARROW-based optofluidic chips. However, more advances can originate from the use of optofluidic approaches to realize multifunctional photonic device and microfluidic elements on the same chip, as it will be illustrated in the following.

4.2. Interferometric Optofluidic Devices for Sensing Applications

Multimode and single-mode liquid ARROWs have been demonstrated as very promising tools for sensing applications because of their high compactness, sensitivity, and the inherit ability to transport the light and the analyte in the same physical channel. The potential for the realization of various functional optical elements on chip makes ARROW a suitable candidate for integration into compact photonic devices. Additionally, simply using the waveguide itself as optofluidic sensor, more sophisticated device architectures can, in principle, be realized by using liquid core ARROWs. In particular, the demonstration of single-mode ARROWs has provided the opportunity to implement optical sensors which make use of interferometric phenomena. For instance multimode interference (MMI) splitters based on liquid ARROWs have been demonstrated [31] and implemented in on-chip optofluidic interferometers [32]. Moreover, curved ARROW sections can be used for the realization of optical devices comprising light splitters or combiners, as it occurs in integrated optical interferometers. Bend waveguides based on ARROWs have been modeled by using 1-D FDM method [16] and fabricated in ring resonator geometry [33].

Planar optical integrated interferometric devices are very attractive for sensing applications as they combine high sensitivity with compactness. Typically, these devices are based on conventional solid core waveguides and exploit evanescent sensing to detect the presence of analytes in fluid sample [34,35,36]. Recently, innovative optofluidic architectures have been proposed for these devices to simplify fluids sampling and manipulation via microfluidic integration [4]. Very interesting optofluidic architectures for interferometric devices have been proposed based on droplet microfluidics [37] and fluid-air interface (meniscus) [38,39]. Several integrated optofluidic interferometric devices have been demonstrated based on liquid ARROWs, like wavelength division multiplexing (WDM) using MMI [40], ring resonators [41], and Mach–Zehnder interferometers (MZIs) [32,42].

Optofluidic schemes for integrated MZIs in which a microfluidic channel is inserted along the sensing arm have been fabricated to allow the direct coupling of the light with sample fluid [43,44]. This design strategy enables a higher sensitivity thanks to the almost complete overlap of the mode power with the sample fluid in the sensing region. However, in these cases the interaction length is limited by the width of the microfluidic channel. Moreover, the high insertion losses at solid/liquid interface in the inserting region can cause poor device performances. With the advantages offered by liquid ARROW, a planar integrated optofluidic MZI entirely shaped by liquid microchannels has been fabricated [42]. The design consists of two asymmetric single-mode branches split from a Y-Branch and recombined by an inverse Y-Branch in single mode waveguide for interfering. Further improvement in sensor design has been proposed in the work of Testa *et al.* [32] to minimize the power unbalancing between arms and to increase the visibility of the interferometer (Figure 6a). The device features large RI tunability, typical of the optofluidic microsystems, as the guiding condition is satisfied for quite a large RI range $n_c = 1.32 \div 1.45$, that corresponds to the detection range of the sensing device. RI limit of detection (LOD) of $\Delta n = 1.6 \times 10^{-5}$ RIU has been estimated, which compares well with results reported for other optofluidic MZI for label-free sensing of liquids [43,45]. However, while evanescent based-MZIs have been demonstrated as extremely sensitive in direct and real-time measurement of biomolecular interactions [46], optofluidic MZIs have been mostly applied for RI volume sensing of liquids due to the enhanced bulk interaction occurring in the optofluidic channel. Detection of single viruses and the discrimination between different kinds of virus types in a sample mixture have been demonstrated using an optofluidic heterodyne interferometer employing an out-of-plane detection/ interrogation optical scheme [47].

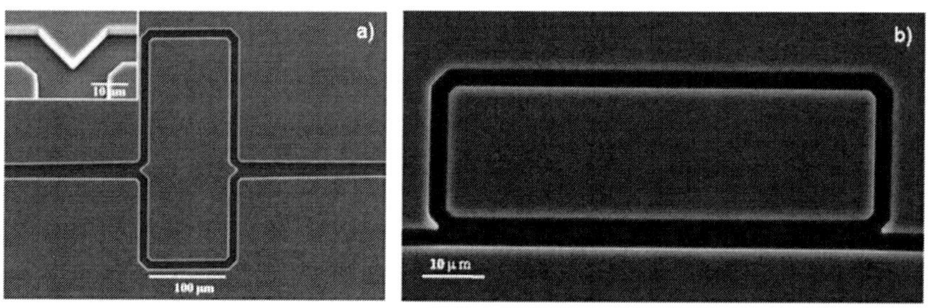

Figure 6. (**a**) SEM image of an integrated MZI based on liquid ARROWs. Reprinted with permission from [32]. Copyright (2010) OSA. (**b**) SEM image of an integrated optofluidic ring resonator based on liquid ARROWs. Reprinted with permission from [41]. Copyright (2010) AIP.

Optofluidic ring resonators (ORRs) have also been recently proposed. Compared with MZI-based sensors that exploit limited sensing lengths, optical resonators benefit from the repetitive interaction between the light and the sample, achieved thanks to the resonant recirculation of light in the cavity. Thanks to this property, highly compact devices can be fabricated with very high sensing capability. Several architectures for optofluidic resonators have been demonstrated which incorporate fluidic capabilities into the photonic structure, such as the capillary-based optofluidic ring resonators (CORRs) [48,49,50]. The CORR is a microfluidic capillary that supports resonant optical modes (the so-called whispering gallery modes (WGMs)) circulating in the capillary wall with the evanescent field extending into the fluid.

Liquid core ARROW has been successfully employed as a basic element to form an integrated ORR (Figure 6b) [41]. A first generation of ARROW-based ORR has been fabricated with a moderate quality factor (Q-factor) on the order of 10^3, mainly due to the high optical losses of the curved sections. Advances in the ORR performances were experimentally obtained by performing an accurate optimization on the initial structure design (first-generation ORR), including those concerning the optofluidic level of integration [16]. A planar ORR based on liquid ARROWs with a Q-factor of about 10^4 has been demonstrated by applying the optimization procedure [33]. The proposed approach retained the merit of high level of optofluidic integration in a planar configuration. The strong light-matter interaction occurring in the liquid core confers high bulk RI sensitivity to the ORR, experimentally estimated as about 700 nm/RIU. This value is larger than that obtained by most of the other optofluidic ring resonator sensors like slot- and capillary-based ORRs [50,51,52]. In order to be applied for biomolecules detection, the surface sensing performances of the device have been simulated, showing very promising capabilities [16]. In comparison, the CORRs have demonstrated very good surface sensing. Biotin molecules captured on the wall surface have been detected with a detection limit of approximately 1.6 pg/mm^2 [51]. Breast cancer biomarkers (CA15-3) have been also detected at a concentration of 1–200 units/mL in phosphate-buffered saline buffer and 20–2500 units/mL in fetal calf serum [53]. However, tapered optical fibers are typically used to excite WGMs, which hinders the integration of CORRs onto a compact lab-on-chip platform.

4.3. Planar Integrated ARROW Platform

One of the most relevant problems associated with the chip-scale integration of liquid core waveguides concerns the way light is collected from the open-ended liquid core. An optimized transmission between liquid core waveguides and off-chip components is necessary to increase the overall optical throughput. In particular, the ability to efficiently collect the light is crucial for achieving single-particle detection. Not surprisingly, the first demonstration of single molecule sensitivity using an open-ended liquid ARROW required bulk optical components, such as a high numerical aperture objective lens [23,30].

Technological solutions to provide on chip integration of intersecting microfluidic channels and optical waveguides are an object of many works in the literature [54,55,56,57,58]. In addition to the obvious implications to enable miniaturization and simplify the alignment procedure, on-chip integrated waveguides offer two further remarkable advantages. First, the possibility to improve the optical coupling at the waveguide-fluid interface that in turn improves the signal to noise ratio and the sensitivity of detection. Second, they can be suitably placed on-chip in order to illuminate only specific points along the liquid core waveguide, thus enabling highly spatially-resolved optical detection on a sub-picoliter excitation volume. Using surface micromachining technology, the first step towards the realization of an integrated platform based on liquid ARROW has been proposed in the work of Yin *et al.* [59]. The authors demonstrated an important advance in sensor design, obtained by integrating solid core with liquid core ARROW in a full planar geometry. Parallel arrays of solid core waveguides were arranged orthogonally with liquid waveguides for fluorescence measurements on sub-picoliter volumes. The proposed approaches resulted in a network of intersecting waveguides for specific excitation of well-defined locations along the liquid core waveguide. Thanks to this geometry, strong pump suppression and sensor sensitivity down to a single molecule have been achieved. Using a similar approach, a solid core waveguide aligned with the liquid waveguide has been integrated on an ARROW platform for surface-enhanced Raman spectroscopy (SERS) detection with molecular specificity [60]. In this configuration the solid waveguide was collinear with the liquid waveguide and it was used to excite the entire liquid volume. Moreover, in order to prevent liquid evaporation during measurements, a further remarkable improvement has been proposed, consisting in an on-chip reservoir of 10 μL volume for continuous piping of fluid solution. This has been the first implementation of a reliable optofluidic tool on an ARROW sensing platform. Using this platform, the authors detected the minimum active R6G concentration of 30 nM, corresponding to measuring approximately 4×10^5 molecules in the ARROW volume of 44 pL. Other very interesting optofluidic devices for SERS detection have been demonstrated with better limits of detections, but generally requiring out-of-plane bulk instruments for the measurements [61,62].

Step-by-step, successive improvements of the optical and sensing performances of the liquid ARROWs have also been made from a fabrication point of view [63,64,65,66]. In [65] tantalum oxide and low temperature-deposited silicon nitride were proposed to reduce the background contribution of native photoluminescence of the cladding material in fluorescence measurements. With this arrangement the authors demonstrated a signal-to-noise-ratio (SNR) improvement by a factor of 12 in fluorescent nanoparticle detection compared to conventional silicon nitride-based ARROWs. Optimization of interface transmission between the liquid and solid ARROWs has also been performed to achieve high sensitivity in spectroscopic measurements [66,67].

A chip-scale integrated planar sensing system that combines optical and fluidic functions is required to address the challenge of high sensitivity and compactness. By using a liquid ARROW integrated with both orthogonal and aligned solid waveguides, together with sample reservoirs, the first planar optofluidic chip for single particle detection and analysis has been demonstrated [68]. The chip has been fabricated by using surface micromachining planar silicon technology. In this design, solid waveguides are not only used to precisely excite a sub-picoliter liquid volume but also to collect and deliver light from liquid ARROW towards off-chip fibers (Figure 7). The advantages offered by this design of providing multiple and spatially separated collection paths have enabled the implementation of advanced multi-color spectroscopy techniques on chip, such as fluorescence cross-correlation spectroscopy (FCCS) for the detection and discrimination between nanobeads labeled with different fluorescent dyes [69]. The chip has been successfully employed to detect a single bacteriophage Qβ virus without the need for particle immobilization [70]. The achieved sensitivity is approaching detection of virus on single level and is comparable to sensitivity reported with other microfluidic chips [71,72,73] and other integrated techniques, such as evanescent waveguide sensors [74,75,76], electrical nanowire arrays [77], and nanoelectromechanical cantilevers [78]. More recently, a solid-state nanopore has been integrated in the optofluidic chip. The nanopore ensures controlled delivery of individual bioparticles into the ARROW optofluidic channels. The bioparticle detection involved both electrical and optical measurements. Using this arrangement, the authors demonstrated single-particle detection of a fluorescently-labeled influenza virus and λ-DNA [79,80]. Optical detection techniques that complement the electrical measurements are well-known as very powerful, high-sensitive tools for single-particle detection using nanopore sensing [81]. However, the proposed optofluidic chip presented the advantage of multi-level integration of fluidic, optical, and electrical control on a single chip.

Manipulation of bio-particles in a fluid sample is of great interest in the field of optofluidics as it enables analysis of biomolecules at a single level [82]. Basically, by using the chip design proposed in [68], very interesting operations have been demonstrated based on an all-optical particle control via the integrated waveguides, *i.e.*, the trapping and sorting of the particles in a fluid flow [83,84,85,86]. The planar nature of the optofluidic chip suggest that this technique could be applied for sample processing of a liquid with suspended bioparticles (viruses, bacteria, cells, and other microorganisms) for on-chip single particles for optical sensing and analysis. The authors exploited the combination of optical trapping and fluorescence detection of single microorganisms by evaluating the photobleaching dynamics of stained DNA in *Escherichia coli* bacteria [85].

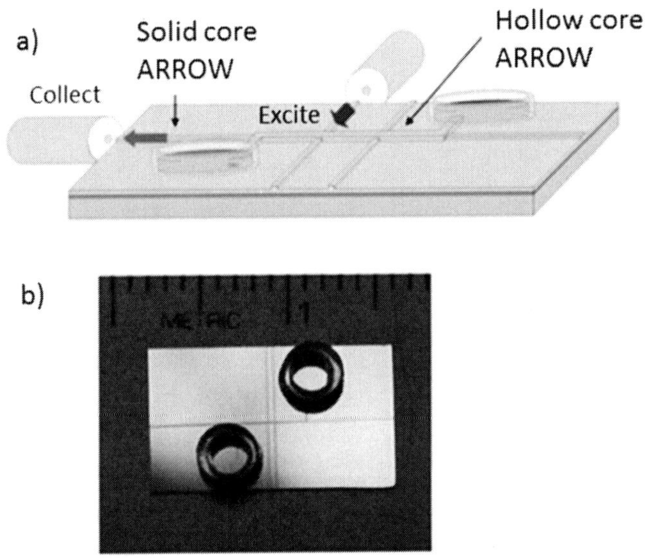

Figure 7. Planar optofluidic chip. (**a**) Schematic view of the integrated optofluidic chip showing hollow- and solid-core ARROW waveguides and (**b**) photograph of fabricated chip. Reprinted with permission from [68]. Copyright (2007) RSC Publishing.

In striving for miniaturization, several efforts have focused on the integration of various optical components on chip, such as slits and lenses [87,88], mirrors [89], filters [90,91], *etc.* Spectral filters are commonly used to reject pump contribution and distinguish the sample signal from background noise. Integration of optical filters on planar platforms can greatly support device miniaturization for high sensitivity in spectroscopic techniques, avoiding the use of bulky optical components [90]. In an ARROW waveguide, wavelength filtering can be achieved by exploiting the inherent spectral dependence of the confinement mechanism [11]. In the work of Philips *et al.* [92], an integrated notch filter has been implemented on a planar ARROW optofluidic chip for spectroscopic applications. The interference claddings of the collection solid waveguides have been suitably designed in order to confine the analyte signal wavelengths while rejecting the excitation light. In the work of Measor *et al.* [93] a similar approach has been used to tailor the spectral response of both liquid and solid core ARROWs. The authors demonstrated the ability to simultaneously reject the pump light while preserving low loss propagation of the analyte signal in Forster resonance energy transfer (FRET) detection of doubly-labeled oligonucleotides. High extinction of 37 dB was shown with a 4 mm long optofluidic filter. The estimated LOD of 15 nM is comparable with other integrated approaches [91].

In the examples presented so far, although the high level of optofluidic integration, the microfluidic functionality is essentially restricted to the ARROW ability to host fluids. On-chip devices for complete bio-medical and chemical analysis, however, often require more sophisticated microfluidic functions, *i.e.*, to accurately mix fluids and control their flow rates. Optofluidics arises from the need to move these functionalities on-chip without sacrificing the optical part. Actually, optofluidics attempts to find design solutions to suitably merge fluidic and photonic elements and improve the overall device functionalities. It is also becoming even more recognized that microfluidic systems can be more easily fabricated using polymer materials and related fabrication technologies. This choice benefits from the reduced cost and simpler fabrication procedures, which could promote larger-scale diffusion. On the basis of these concerns, ARROW-based optofluidic chips that make use of a hybrid technology to integrate microfluidic system have been recently developed by exploiting both surface and bulk micromachining fabrication procedures.

Based on bulk micromachining process, a hybrid ARROW optofluidic platform has been developed using the combination of PDMS and silicon-based materials. PDMS is extensively used in microfluidic technology because of its optical transparency in the visible/UV region, the biocompatibility and the low-cost fabrication procedures, like soft lithography [94]. The opportunity of combining silicon and polymer technologies has been firstly demonstrated with the fabrication of a single-mode h-ARROW, as reported in the work of Testa *et al.* [28]. Starting from this vision, the optofluidic platform has been assembled in a modular structure, with the upper polymer parts having both optical and microfluidic functions [95] (Figure 8). In particular, by moving from the bottom to the top of the structure, in the first PDMS layer (layer a) hybrid solid-core ARROWs have been integrated to deliver the optical signal towards the collection optical fibers. They also sealed the open-ended channels, preventing liquid evaporation. Merely fluid handling functionalities were restricted to the second PDMS layer (layer b), where a microfluidic mixer was incorporated. Such layer can be easily replaced with another, in order to adapt the microfluidic system to specific sensing applications. This confers to the platform high design flexibility. A multimode h-ARROW was used in order to allow self-aligned positioning of off-chip fibers with solid waveguides through the optofluidic channels. The sensing performance of the device was tested by performing fluorescence measurements at different dye concentrations, controlled by means of the integrated micromixer; an LOD of 2.5 nM was demonstrated. Even if this value is high as compared with the current state of the art of fluorescence measurement using evanescent waveguide sensors [96], the proposed modular approach for chip assembling into a compact and multifunctional format is very promising and offers prospects of an increased flexibility. For comparison, a modular approach for slot biochemical sensors have also been recently proposed in the work of

Carlborg *et al.* [52] An array of optical slot-waveguide ring resonator sensors has been packaged into a chip incorporating microfluidic sample handling, with a microfluidic layer assembled in a compact cartridge. The authors demonstrated a volume refractive index detection limit of 5×10^{-6} RIU and surface mass detection limit of 0.9 pg/mm^2.

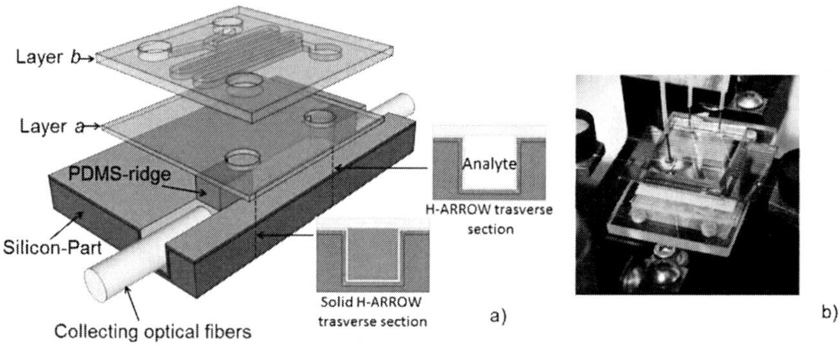

Figure 8. (**a**) Schematic of the proposed hybrid ARROW optofluidic platform with transverse section of liquid h-ARROW and solid h-ARROW; (**b**) Photograph of fabricated chip. Reprinted with permission from [95]. Copyright (2014) OSA.

A hybrid ARROW optofluidic platform has been also proposed in the work of Parks *et al.* [97]. In this configuration, the fluidic and optical chips were two separate entities, connected via metal reservoirs as shown in Figure 9a. The fluidic layer was fabricated in PDMS by soft lithography, being aligned but not in contact with the optical layer. This allows great design reconfigurability, as the PDMS layer can be removed and replaced without losing the alignment of the fluidic reservoirs with the optical chip. Microfluidic capabilities of fluidic mixing, distribution, and filtering have been demonstrated in combination with single-molecule fluorescent detection of dye-labeled λ-DNA. Using a similar approach for microfluidic integration, a more sophisticated fluid handling system capable of advanced microfluidic functions has been integrated with the optofluidic chip (Figure 9b). In particular, an actively programmable microfluidic layer made of PDMS was obtained by employing an interconnected array of nanoliter microvalves. Precise sample handling functions such as mixing, splitting, delivering, and storing were implemented for molecular diagnosis of viral nucleic acids from complex solutions [98], and for amplification-free detection and quantification of Ebola virus on clinical samples [99]. By using the advantages of the integrated PDMS-based microfluidic chip (automaton) to process the fluid sample, the authors demonstrated a LOD of 0.2 pfu/mL, comparable with other amplification-free methods like PCR analysis [100] and more than four orders of magnitude lower than other chip-based approaches [101].

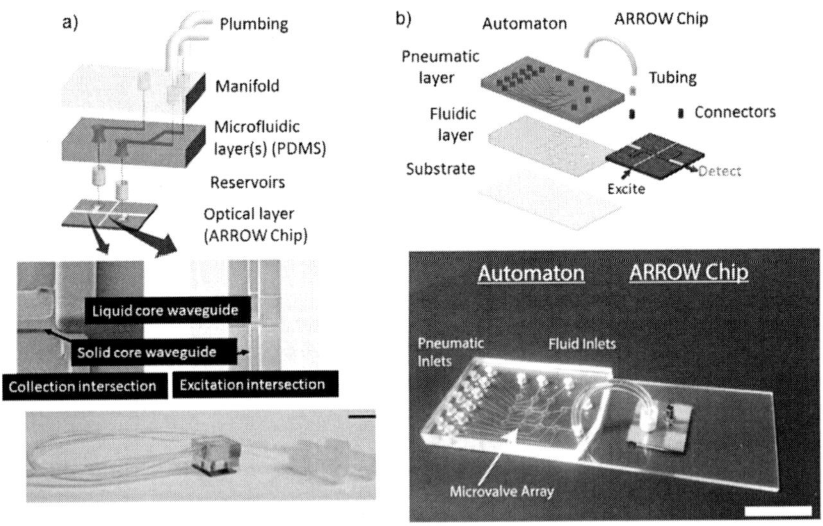

Figure 9. (**a**) Expanded view of PDMS integration with an ARROW optofluidic chip. Pictures are SEM images of optical intersections. Reprinted with permission from [97]. Copyright (2013) RSC Publishing. (**b**) Schematic of hybrid integration of automaton and ARROW chips. Reprinted with permission from [98]. Copyright (2014) AIP Publishing.

5. CONCLUSIONS

The discussion in this review demonstrates that the liquid core ARROW is a valuable tool for optofluidic integration. ARROWs have been successfully fabricated using current silicon technologies, showing very high optical performances. With core sections of a few microns, these waveguides can contain ultra-small fluid volumes of nanoliters to a few picoliters and simultaneously analyze it down to the single-molecule detection. As demonstrated by discussed examples, ARROWs have the capability to enhance the sensitivity of detection in several optical methods of analysis like fluorescence, Raman spectroscopy, and refractometry. Using ARROWs, complex interferometric sensing devices like a Mach–Zehnder interferometer or ring resonators have been designed and successfully tested.

Integrated optofluidic platforms have been demonstrated by combining solid core with liquid core ARROWs and microfluidic devices in several planar configurations. This represents a significant step forward from the previous version of an ARROW platform, opening a path towards an effective realization of chip-scale integrated system. The integration of planar optical components with microfluidic tools at the chip-scale represents the future trend of sensing systems for point-of care diagnosis or *in situ* monitoring. Increasing the synergy between these technologies can lead to highly reliable and portable systems.

By using ARROW optofluidic platforms, several advanced sensing methods have been successfully implemented for the manipulation, detection and analysis of single bioparticles directly on chip. The integration of microfluidic systems capable of advanced operations into an ARROW optofluidic chip has been the topic of intense investigation in recent years. Novel solutions have been proposed that are based on hybrid silicon/polymer technologies. The use of polymer to fabricate the microfluidic system has the potential to address the requirements of reduced fabrication cost and simplified manufacturing. The possibility to realize an optofluidic infrastructure of interconnected solid- and liquid-core waveguides on a planar silicon substrate and the hybrid integration of the ARROW platform has been demonstrated to be very attractive and creates a pathway for further integration of optoelectronic components towards a complete and self-contained chip integrated system. Based on the great improvements obtained in recent years, further progresses can be envisioned in fabrication methods that will improve the overall functionalities and promote commercial potential of ARROW-based optofluidic chip.

ACKNOWLEDGMENTS

The research leading to these results has received the financial support of the Italian Minister of University and Research (MIUR), Futuro in Ricerca (FIR) programme under the grant N. RBFR122KL1 (SENS4BIO).

CONFLICTS OF INTEREST

The authors declare no conflict of interest.

REFERENCES

1. Monat, C.; Domachuk, P.; Eggleton, B.J. Integrated optofluidics: A new river of light. Nat. Photon. 2007, 1, 106–114.

2. Psaltis, D.; Quake, S.R.; Yang, C. Developing optofluidic technology through the fusion of microfluidics and optics.Nature 2006, 442, 381–386.

3. Fan, X.; White, I.M. Optofluidic microsystems for chemical and biological analysis. Nat. Photon. 2011, 5, 591–597.

4. Testa, G.; Persichetti, G.; Bernini, R. Optofluidic Approaches for Enhanced Microsensor Performances. Sensors 2015, 15, 465–484.

5. Schmidt, H.; Hawkins, A.R. Optofluidic waveguides: I. Concepts and implementations. Microfluid. Nanofluid. 2008, 4, 3–16.

6. Bernini, R.; Campopiano, S.; Zeni, L. Silicon micromachined hollow optical waveguides for sensing applications. IEEE J. Sel. Top. Quant. Electron. 2002, 8, 106–110.

7. Erickson, D.; Rockwood, T.; Emery, T.; Scherer, A.; Psaltis, D. Nanofluidic tuning of photonic crystal circuits. Opt. Lett.2006, 31, 59–61.

8. Surdo, S.; Merlo, S.; Carpignano, F.; Strambini, L.M.; Trono, C.; Giannetti, A.; Baldini, F.; Barillaro, G. Optofluidic microsystems with integrated vertical one-dimensional photonic crystals for chemical analysis. Lab Chip 2012, 12, 4403–4415.

9. Almeida, V.R.; Xu, Q.; Barrios, C.A.; Lipson, M. Guiding and confining light in void nanostructure. Opt. Lett. 2004, 29, 1209–1211.

10. Barrios, C.A. Optical Slot-Waveguide Based Biochemical Sensors. Sensors 2009, 9, 4751–4765.

11. Schmidt, H.; Yin, D.; Barber, J.P.; Hawkins, A.R. Hollow-core waveguides and 2-D waveguide arrays for integrated optics of gases and liquids. IEEE J. Sel. Top. Quant. Electron. 2005, 11, 519–527.

12. Delonge, T.; Fouckhardt, H. Integrated optical detection cell based on Bragg reflecting waveguides. J. Chromatogr. A1995, 716, 135–139.

13. Baba, T.; Kokubun, Y. Dispersion and radiation loss characteristics of antiresonant reflecting optical waveguides –

numerical results and analytical expressions. IEEE J. Quantum. Electron. 1992, 28, 1689–1700.

14. Litchinitser, N.M.; Abeeluck, A.K.; Headley, C.; Eggleton, B.J. Antiresonant reflecting photonic crystal optical waveguides. Opt. Lett. 2002, 27, 1592–1594.

15. Archambault, J.-L.; Black, R.J.; Lacroix, S.; Bures, J. Loss calculations for antiresonant waveguides. J. Lightwave Technol. 1993, 11, 416–423.

16. Testa, G.; Persichetti, G.; Bernini, R. Design and optimization of an optofluidic ring resonator based on liquid-core hybrid ARROWs. IEEE Photon. J. 2014, 6, 2201614.

17. Bernini, R.; De Nuccio, E.; Minardo, A.; Zeni, L.; Sarro, P.M. Integrated silicon optical sensors based on hollow core waveguide. Proc. SPIE 2007, 6477, 647714.

18. Gollub, A.H.; Carvalho, D.O.; Paiva, T.C.; Alayo, M.I. Hollow core ARROW waveguides fabricated with SiOxNy films deposited at low temperatures. Phys. Status Solidi C 2010, 7, 964–967.

19. Yin, D.; Deamer, D.W.; Schmidt, H.; Barber, J.P.; Hawkins, A.R. Integrated optical waveguides with liquid cores. Appl. Phys. Lett. 2004, 85, 3477.

20. Holmes, M.; Keeley, J.; Hurd, K.; Schmidt, H.; Hawkins, A.R. Optimized pirhana etching process for SU8-based MEMS and MOEMS construction. J. Micromech. Microeng. 2010, 20, 115008.

21. Yin, D.; Schmidt, H.; Barber, J.P.; Hawkins, A.R. Integrated ARROW waveguides with hollow cores. Opt. Express 2004,12, 2710–2715.

22. Yin, D.; Barber, J.P.; Hawkins, A.R.; Schmidt, H. Waveguide loss optimization in hollow-core ARROW waveguides.Opt. Express 2005, 13, 9331–9336.

23. Yin, D.; Barber, J.P.; Hawkins, A.R.; Schmidt, H. Highly efficient fluorescence detection in picoliter volume liquid-core waveguides. Appl. Phys. Lett. 2005, 87, 211111.

24. Testa, G.; Huang, Y.; Zeni, L.; Sarro, P.M.; Bernini, R. Liquid core ARROW waveguides by atomic layer deposition.IEEE Photon. Technol. Lett. 2010, 22, 616–618.

25. Campopiano, S.; Bernini, R.; Zeni, L.; Sarro, P.M. Microfluidic sensor based on integrated optical hollow waveguides.Opt. Lett. 2004, 29, 1894–1896.

26. Barber, J.P.; Lunt, E.J.; George, Z.A.; Yin, D.; Schmidt, H.; Hawkins, A.R. Integrated hollow waveguide with arc shaped core. IEEE Photon. Technol. Lett. 2006, 18, 28–30.

27. Yin, D.; Schmidt, H.; Barber, J.P.; Lunt, E.J.; Hawkins, A.R. Optical characterization of arch-shaped ARROW waveguides with liquid cores. Opt. Express 2005, 13, 10564–10570.

28. Testa, G.; Huang, Y.; Zeni, L.; Sarro, P.M.; Bernini, R. Hybrid Silicon-PDMS optofluidic ARROW waveguide. IEEE Photon. Technol. Lett. 2012, 24, 1307–1309.

29. Bernini, R.; De Nuccio, E.; Brescia, F.; Minardo, A.; Zeni, L.; Sarro, P.M.; Palumbo, R.; Scarfi, M.R. Development and characterization of an integrated silicon micro flow cytometer. Anal. BioAnal. Chem. 2006, 386, 1267–1272.

30. Measor, P.; Seballos, L.; Yin, D.; Barber, J.P.; Hawkins, A.R.; Zhang, J.; Schmidt, H. Integrated ARROW waveguides for molecule specific surface-enhanced Raman sensing. In Proceedings of the 2006 16th Biennial University/Government/Industry Microelectronics Symposium, San Jose, CA, USA, 25–28 June 2006; pp. 181–182.

31. Bernini, R.; Testa, G.; Zeni, L.; Sarro, P.M. A 2 × 2 Optofluidic multimode interference coupler. IEEE J. Sel. Top. Quantum Electron. 2009, 15, 1478–1484.

32. Testa, G.; Huang, Y.; Sarro, P.M.; Zeni, L.; Bernini, R. High-visibility optofluidic Mach–Zehnder interferometer. Opt. Lett. 2010, 35, 1584–1586.

33. Testa, G.; Collini, C.; Lorenzelli, L.; Bernini, R. Planar silicon-polydimethylsiloxane optofluidic ring resonator sensors.IEEE Photon. Tech. Lett. 2016, 28, 155–158.

34. Estevez, M.-C.; Alvarez, M.; Lechuga, L.M. Integrated optical devices for lab-on-a-chip biosensing applications. Laser Photonics Rev. 2012, 6, 463–487.

35. Prieto, F.; Sepúlveda, B.; Calle, A.; Llobera, A.; Domínguez, C.; Abad, A.; Montoya, A.; Lechuga, L.M. An integrated optical interferometric nanodevice based on silicon technology for biosensor applications. Nanotechnology 2003, 14, 907–912.

36. Nitkowski, A.; Baeumner, A.; Lipson, M. On-chip spectrophotometry for bioanalysis using microring resonators. Biomed. Opt. Express 2011, 2, 271–277.

37. Chin, L.K.; Liu, A.Q.; Soh, Y.C.; Lim, C.S.; Lin, C.L. A reconfigurable optofluidic Michelson interferometer using tunable droplet grating. Lab Chip 2010, 10, 1072–1078.

38. Grillet, C.; Domachuk, P.; Ta'eed, V.; Mägi, E.; Bolger, J.A.; Eggleton, B.J.; Rodd, L.E.; Cooper-White, J. Compact tunable microfluidic interferometer. Opt. Express 2004, 12, 5440–5447.

39. Domachuk, P.; Grillet, C.; Ta'eed, V.; Mägi, E.; Bolger, J.; Eggleton, B.J.; Rodd, L.E.; Cooper-White, J. Microfluidic interferometer. Appl. Phys. Lett. 2005, 86, 024103.

40. Ozcelik, D.; Parks, J.W.; Wall, T.A.; Stott, M.A.; Cai, H.; Parks, J.W.; Hawkins, A.R.; Schmidt, H. Optofluidic wavelength division multiplexing for single-virus detection. Proc. Natl. Acad. Sci. USA 2015, 112, 12933–12937.

41. Testa, G.; Huang, Y.; Sarro, P.M.; Zeni, L.; Bernini, R. Integrated optofluidic ring resonator. Appl. Phys. Lett. 2010, 97, 131110.

42. Bernini, R.; Testa, G.; Zeni, L.; Sarro, P.M. Integrated optofluidic Mach–Zehnder interferometer based on liquid core waveguides. Appl. Phys. Lett. 2008, 93, 011106.

43. Dumais, P.; Callender, C.L.; Noad, J.P.; Ledderhof, C.J. Integrated optical sensor using a liquid-core waveguide in a Mach–Zehnder interferometer. Opt. Express 2008, 16, 18164–18172.

44. Crespi, A.; Gu, Y.; Ngamsom, B.; Hoekstra, H.J.W.M.; Dongre, C.; Pollnau, M.; Ramponi, R.; van den Vlekkert, H.H.; Watts, P.; Cerullo, G.; et al. Three-dimensional Mach–Zehnder interferometer in a microfluidic chip for spatially-resolved label-free detection. Lab Chip 2010, 10, 1167–1173.

45. Lapsley, M.I.; Chiang, I.-K.; Zheng, Y.B.; Ding, X.; Mao, X.; Huang, T.J. A single-layer, planar, optofluidic Mach–Zehnder interferometer for label-free detection. Lab Chip 2011, 11, 1795–1800.

46. Kozma, P.; Kehl, F.; Ehrentreich-Förster, E.; Stamm, C.; Bier, F.F. Integrated planar optical waveguide interferometer biosensors: A comparative review. Biosens. Bioelectron. 2014, 58, 287–307.

47. Mitra, A.; Deutsch, B.; Ignatovich, F.; Dykes, C.; Novotny, L. Nano-optofluidic Detection of Single Viruses and Nanoparticles. ACS Nano 2010, 4, 1305–1312.

48. Huang, G.; Bolaños Quiñones, V.A.; Ding, F.; Kiravittaya, S.; Mei, Y.; Schmidt, O.G. Rolled-up optical microcavities with

subwavelength wall thicknesses for enhanced liquid sensing applications. ACS Nano 2010, 4, 3123–3130.

49. Berneschi, S.; Farnesi, D.; Cosi, F.; Conti, G.N.; Pelli, S.; Righini, G.C.; Soria, S. High Q silica microbubble resonators fabricated by arc discharge. Opt. Lett. 2011, 36, 3521–3523.

50. White, I.M.; Oveys, H.; Fan, X. Liquid-core optical ring-resonator sensors. Opt. Lett. 2006, 31, 1319–1321.

51. Li, H.; Fan, X. Characterization of sensing capability of optofluidic ring resonator biosensors. Appl. Phys. Lett. 2010, 97, 011105.

52. Carlborg, C.F.; Gylfason, K.B.; Kaźmierczak, A.; Dortu, F.; Bañuls Polo, M.J.; Maquieira Catala, A.; Kresbach, G.M.; Sohlström, H.; Moh, T.; Vivien, L.; et al. A packaged optical slot-waveguide ring resonator sensor array for multiplex label-free assays in labs-on-chips. Lab Chip 2010, 10, 281–290.

53. Zhu, H.; Dale, P.S.; Caldwell, C.W.; Fan, X. Rapid and Label-Free Detection of Breast Cancer Biomarker CA15-3 in Clinical Human Serum Samples with Optofluidic Ring Resonator Sensors. Anal. Chem. 2009, 81, 9858–9865.

54. Fei, P.; Chen, Z.; Men, Y.; Li, A.; Shen, Y.; Huang, Y. A compact optofluidic cytometer with integrated liquid-core/PDMS-cladding waveguides. Lab Chip 2012, 12, 3700–3706.

55. Fleger, M.; Neyer, A. PDMS microfluidic chip with integrated waveguides for optical detection. Microelectron. Eng.2006, 83, 1291–1293.

56. Mogensen, K.B.; El-Ali, J.; Wolff, A.; Kutter, J.P. Integration of polymer waveguides for optical detection in microfabricated chemical analysis systems. Appl. Opt. 2003, 42, 4072–4079.

57. Friis, P.; Hoppe, K.; Leistiko, O.; Mogensen, K.B.; Hübner, J.; Kutter, J.P. Monolithic integration of microfluidic channels and

optical waveguides in silica on silicon. Appl. Opt. 2001, 40, 6246–6251.

58. Wang, Z.; Yan, H.; Wang, Z.; Zou, Y.; Yang, C.-J.; Chakravarty, S.; Subbaraman, H.; Tang, N.; Xu, X.; Fan, D.L.; et al. Ultralow-loss waveguide crossings for the integration of microfluidics and optical waveguide sensors. Proc. SPIE2015, 9320, 932012.

59. Yin, D.; Deamer, D.W.; Schmidt, H.; Barber, J.P.; Hawkins, A.R. Single-molecule detection sensitivity using planar integrated optics on a chip. Opt. Lett. 2006, 31, 2136–2138.

60. Measor, P.; Seballos, L.; Yin, D.; Zhang, J.Z.; Lunt, E.J.; Hawkins, A.R.; Schmidt, H. On-chip surface-enhanced Raman scattering detection using integrated liquid-core waveguides. Appl. Phys. Lett. 2007, 90, 211107.

61. Li, Q.-L.; Li, B.-W.; Wang, Y.-Q. Surface-enhanced Raman scattering microfluidic sensor. RSC Adv. 2013, 3, 13015–13026.

62. Wang, M.; Jing, N.; Chou, I.-H.; Cote, G.L.; Kameoka, J. An optofluidic device for surface enhanced Raman spectroscopy. Lab Chip 2007, 7, 630–632.

63. Lunt, E.J.; Phillips, B.S.; Jones, C.J.; Hawkins, A.R.; Measor, P.; Kuehn, S.; Schmidt, H. Hollow waveguide optimization for fluorescence based detection. Proc. SPIE 2008, 6883, 68830H.

64. Lunt, E.J.; Wu, B.; Keeley, J.M.; Measor, P.; Schmidt, H.; Hawkins, A.R. Hollow ARROW Waveguides on Self-Aligned Pedestals for Improved Geometry and Transmission. IEEE Photonics Technol. Lett. 2010, 22, 1147–1149.

65. Zhao, Y.; Jenkins, M.H.; Measor, P.; Leake, K.; Liu, S.; Schmidt, H.; Hawkins, A.R. Hollow waveguides with low intrinsic photoluminescence fabricated with Ta2O5 and SiO2 film. Appl. Phys. Lett. 2011, 98, 091104.

66. Phillips, B.S.; Jenkins, M.H.; Liu, S.; Schmidt, H.; Hawkins, A.R. Selective Thin-Film deposition for optofluidic platforms with optimized transmission. IEEE Photon. Technol. Lett. 2011, 23, 721–723.

67. Zhao, Y.; Leake, K.D.; Measor, P.; Jenkins, M.H.; Keeley, J.; Schmidt, H.; Hawkins, A.R. Optimization of interface transmission between integrated solid core and optofluidic waveguides. IEEE Photon. Technol. Lett. 2012, 24, 46–48.

68. Yin, D.; Lunt, E.J.; Rudenko, M.I.; Deamer, D.W.; Hawkins, A.R.; Schmidt, H. Planar optofluidic chip for single particle detection, manipulation, and analysis. Lab Chip 2007, 7, 1171–1175.

69. Chen, A.; Eberle, M.M.; Lunt, E.J.; Liu, S.; Leake, K.; Rudenko, M.I.; Hawkins, A.R.; Schmidt, H. Dual-color fluorescence cross-correlation spectroscopy on a planar optofluidic chip. Lab Chip 2011, 11, 1502–1506.

70. Rudenko, M.I.; Kühn, S.; Lunt, E.J.; Deamer, D.W.; Hawkins, A.R.; Schmidt, H. Ultrasensitive Qβ phage analysis using fluorescence correlation spectroscopy on an optofluidic chip. Biosens. Bioelectron. 2009, 24, 3258–3263.

71. Bao, N.; Wang, J.; Lu, C. Recent advances in electric analysis of cells in microfluidic systems. Anal. Bioanal. Chem. 2008,391, 933–942.

72. Choi, J.; Kang, M.; Jung, J.H. Integrated micro-optofluidic platform for real-time detection of airborne microorganisms. Sci. Rep. 2015, 5, 15983.

73. Liu, Y.S.; Walter, T.M.; Chang, W.J.; Lim, K.S.; Yang, L.; Lee, S.W.; Aronson, A.; Bashir, R. Electrical detection of germination of viable model Bacillus anthracis spores in microfluidic biochips. Lab Chip 2007, 7, 603–610.

74. Mukundan, H.; Anderson, A.S.; Grace, W.K.; Grace, K.M.; Hartman, N.; Martinez, J.S.; Swanson, B.I. Waveguide-Based Biosensors for Pathogen Detection. Sensors 2009, 9, 5783–5809.

75. Martinez, J.S.; Grace, W.K.; Grace, K.M.; Hartman, N.; Swanson, B.I. Pathogen detection using single mode planar optical waveguides. J. Mater. Chem. 2005, 15, 4639–4647.

76. Donaldson, K.A.; Kramer, M.F.; Lim, D.V. A rapid detection method for Vaccinia virus, the surrogate for smallpox virus. Biosens. Bioelectron. 2004, 20, 322–327.

77. Patolsky, F.; Zheng, G.; Hayden, O.; Lakadamyali, M.; Zhuang, X.; Lieber, C.M. Electrical detection of single viruses.Proc. Natl. Acad. Sci. USA 2004, 101, 14017–14022.

78. Ilic, B.; Yang, Y.; Craighead, H.G. Virus detection using nanoelectromechanical devices. Appl. Phys. Lett. 2004, 85, 2604.

79. Shuo, L.; Zhao, Y.; Parks, J.W.; Deamer, D.W.; Hawkins, A.R.; Schmidt, H. Correlated electrical and optical analysis of single nanoparticles and biomolecules on a nanopore-gated optofluidic chip. Nano Lett. 2014, 14, 4816–4820.

80. Shuo, L.; Wall, T.A.; Ozcelik, D.; Parks, J.W.; Hawkins, A.R.; Schmidt, H. Electro-optical detection of single λ-DNA.Chem. Comm. 2015, 51, 2084–2087.

81. Gilboa, T.; Meller, A. Optical sensing and analyte manipulation in solid-state nanopores. Analyst 2015, 140, 4733–4747.

82. Kayani, A.A.; Khoshmanesh, K.; Ward, S.A.; Mitchell, A.; Kalantar-zadeh, K. Optofluidics incorporating actively controlled micro- and nano-particles. Biomicrofluidics 2012, 6, 031501.

83. Kühn, S.; Measor, P.; Lunt, E.J.; Phillips, B.S.; Deamer, D.W.; Hawkins, A.R.; Schmidt, H. Loss-based optical trap for on-chip particle analysis. Lab Chip 2009, 9, 2212–2216.

84. Kühn, S.; Lunt, E.J.; Phillips, B.S.; Hawkins, A.R.; Schmidt, H. Optofluidic particle concentration by a long-range dual-beam trap. Opt. Lett. 2009, 34, 2306–2308.

85. Kühn, S.; Phillips, B.S.; Lunt, E.J.; Hawkins, A.R.; Schmidt, H. Ultralow power trapping and fluorescence detection of single particles on an optofluidic chip. Lab Chip 2010, 10, 189–194.

86. Leake, K.D.; Phillips, B.S.; Yuzvinsky, T.D.; Hawkins, A.R.; Schmidt, H. Optical particle sorting on an optofluidic chip.Opt. Express 2013, 21, 32605–32610.

87. Schuhladen, S.; Banerjee, K.; Stürmer, M.; Müller, P.; Wallrabe, U.; Zappe, H. Variable optofluidic slit aperture. Light Sci. Appl. 2016, 5, e16005.

88. Seow, Y.C.; Lim, S.P.; Lee, H.P. Optofluidic variable-focus lenses for light manipulation. Lab Chip 2012, 12, 3810–3815.

89. Müller, P.; Kopp, D.; Llobera, A.; Zappe, H. Optofluidic router based on tunable liquid-liquid mirrors. Lab Chip 2014,14, 737–743.

90. Llobera, A.; Demming, S.; Joensson, H.N.; Vila-Planas, J.; Andersson-Svahn, H.; Büttgenbach, S. Monolithic PDMS passband filters for fluorescence detection. Lab Chip. 2010, 10, 1987–1992.

91. Dandin, M.; Abshire, P.; Smela, E. Optical filtering technologies for integrated fluorescence sensors. Lab Chip 2007, 7, 955–977.

92. Phillips, B.S.; Measor, P.; Zhao, Y.; Schmidt, H.; Hawkins, A.R. Optofluidic notch filter integration by lift-off of thin films. Opt. Express 2010, 18, 4790–4795.

93. Measor, P.; Phillips, B.S.; Chen, A.; Hawkins, A.R.; Schmidt, H. Tailorable integrated optofluidic filters for biomolecular detection. Lab Chip 2011, 11, 899–904.

94. Ng, J.M.; Gitlin, I.; Stroock, A.D.; Whitesides, G.M. Components for integrated poly(dimethylsiloxane) microfluidic systems. Electrophoresis 2002, 23, 3461–3473.

95. Testa, G.; Persichetti, G.; Sarro, P.M.; Bernini, R. A hybrid silicon-PDMS optofluidic platform for sensing applications.Biomed. Opt. Express 2014, 5, 417–426.

96. Taitt, C.R.; Anderson, G.P.; Ligler, F.S. Evanescent wave fluorescence biosensors. Biosens. Bioelectron. 2005, 20, 2470–2487.

97. Parks, J.W.; Cai, H.; Zempoaltecatl, L.; Yuzvinsky, T.D.; Leake, K.; Hawkins, A.R.; Schmidt, H. Hybrid optofluidic integration. Lab Chip 2013, 13, 4118–4123.

98. Parks, J.W.; Olson, M.A.; Kim, J.; Ozcelik, D.; Cai, H.; Carrion, R.; Patterson, J.L.; Mathies, R.A.; Hawkins, A.R.; Schmidt, H. Integration of programmable microfluidics and on-chip fluorescence detection for biosensing applications. Bio-microfluidics 2014, 8, 054111.

99. Cai, H.; Parks, J.W.; Wall, T.A.; Stott, M.A.; Stambaugh, A.; Alfson, K.; Griffiths, A.; Mathies, R.A.; Carrion, R.; Patterson, J.L.; et al. Optofluidic analysis system for amplification-free, direct detection of Ebola infection. Sci. Rep.2015, 5, 14494.

100. Trombley, A.R.; Wachter, L.; Garrison, J.; Buckley-Beason, V.-A.; Jahrling, J.; Hensley, L.E.; Schoepp, R.J.; Norwood, D.A.; Goba, A.; Fair, J.N.; et al. Comprehensive panel of real-time TaqMan™ polymerase chain reaction assays for detection and absolute quantification of Filoviruses, Arenaviruses, and New World Hantaviruses. Am. J. Trop. Med. Hyg. 2010, 82, 954–960.

101. Daaboul, G.G.; Lopez, C.A.; Chinnala, J.; Goldberg, B.B.; Connor, J.H.; Ünlü, M.S. Digital sensing and sizing of vesicular stomatitis virus pseudotypes in complex media: A model for Ebola and Marburg detection. ACS Nano 2014,8, 6047–6055.

6

NUMERICAL STUDY OF HIGHLY NONLINEAR PHOTONIC CRYSTAL FIBER WITH TUNABLE ZERO DISPERSION WAVELENGTHS

Alejandro Barrientos García[1*], Igor A. Sukhoivanov[1], José Amparo Andrade Lucio[1], Oscar Gerardo Ibarra Manzano[1], Igor Guryev[2], Juan Carlos Hernández García[1,3], Gabriel Ramos Ortiz[4]

[1]Department of Electronics Engineering, Universidad de Guanajuato, Salamanca, Mexico

[2]Department of Multidisciplinary Studies, Universidad de Guanajuato, Yuriria, Mexico

[3]CONACYT, Cátedras CONACYT, Ciudad de Mexico, Mexico

[4]Centro de Investigaciones en Óptica, A. C., GPOM, León, Mexico

ABSTRACT

Solid-core silica photonic crystal fiber is proposed borrowing the concept of golden ratio (1.618) and keeping it between pitch and air hole diameter Λ/d in a subset of six rings of air-holes with hexagonal arrangement. In the case when we have a pitch equal to one micron (Λ = 1 μm), we need air-holes diameters d = 0.618 μm in order to achieve two zero dispersion wavelength (ZDW) points at 725 nm and 1055 nm; this gives us the possibility to use the fiber for supercontinuum generation, by pumping close to that points, pulse compression or reshaping. We analyzed a series of fibers using this relation and showed the possibilities of tunable ZDW in a wide range of wavelengths from 725 nm to 2000 nm, with low losses and small effective area. In agreement with the ZDW point needed, the geometry of the structure can be modified to the point of having only three rings of air holes that surround the solid core with low losses and good confinement mode. The design proposed here is analyzed using the finite element method with perfectly matched layers, including the material dispersion directly into the model applying the Sellmeier's equation.

KEYWORDS

Photonic Crystal Fibers, Supercontinuum Generation, Zero Dispersion Wavelength

1. INTRODUCTION

Today the necessities for different applications in science and engineering of photonic crystal fibers (PCF) have boosted proposals on novel designs that satisfy specific conditions of linear and non-linear properties. Because of this, the control of dispersion, losses, mode confinement and conditions for generation of supercontinuum or pulse reshaping is a wide field of research. A series of techniques implying modification of some parameters of the fiber for various applications have been published. Since the invention of PCFs [1] [2] , their designs and applications have been increased. However, some structures suffer from complex geometry. Traditionally optical fibers are used in various applications, for instance, to transport laser energy for material machining, as a gain medium in fiber lasers or optical telecommunication [3]. However, new possibilities of applications arise when nonlinear properties of PCFs are manifested. Specifically, nonlinearities in PCFs allow new light sources with very broad spectra, i.e., supercontinuum generation (SCG) [4] [5] . The possibility to control group velocity dispersion (GVD) in PCFs, their small effective area and elevated nonlinearity [6] has allowed the observation of SCG in a much wider range of source parameters than has been possible with bulk media or conventional fibers [7] . It is known that in the case of normal GVD is present in the PCFs, the Raman scattering effect and self-phase modulation (SPM) dominate the spectrum broadening [8] , and soliton dynamics govern spectral broadening when central wavelength of the propagating pulse is in anomalous GVD. Finite element and finite-difference methods as well as the multipole method and the plane wave expansion method [9] [10] were used to characterize the electromagnetic field distribution in PCFs. The accurate control of the dispersion characteristics is obligatory in variety of applications, such as pulse recompression [11] mode locked fiber lasers [12] , nonlinear effects, including Raman scattering, four-wave mixing, SPM or soliton generation [13] , and SCG [5] [7] . The flexible geometry of PCFs offers many unique and novel dispersion properties in comparison to conventional single mode fibers such as nearly zero ultraflattened dispersion [14] -[16] , high negative chromatic dispersion [17] , and anomalous dispersion in short-wavelength regime [18] . As is known, the chromatic dispersion coefficient of an optical fiber is a key quantity in various analysis and design issues in fiber transmission systems [19] [20] .

According to the guidance mechanism, PCFs can be divided into two categories: first, total internal reflection (TIR), where air holes surround a high index solid core, and second, photonic band gap (PBG), where the light is confined to a central air hole [21] . In PCFs, the cladding region is usually formed by array of air holes in a regular triangular lattice with separation Λ [22] . Although there is a rich nomenclature of PCFs at this moment, we will restrict ourselves to index-guiding PCFs. In this kind of fibers, the core index is greater than the average index of the cladding because of the presence of air holes, and the fiber can guide the light by total internal reflection as a standard fiber does. That is, the guided mode has effective index n_{eff} that satisfies the condition

$$n_{co} > \left(n_{eff} = \frac{\beta}{k_0} \right) > n_{FSM} \tag{1}$$

where β is the mode propagation constant, n_{eff} is the core index, and n_{FSM} is the cladding effective index, which is termed also the effective index of the fundamental space-filling mode (FSM). In case of a PCF made of pure silica, n_{co} is the refractive index of silica. The losses in PCFs occur due to intrinsic material absorption, structural imperfection, Rayleigh scattering, and imperfect confinement. Fabrication-related losses can be reduced by carefully optimizing the fabrication process [23] . Confinement losses are an additional form of energy leakage that occurs in single-material fibers because of inherently leaky nature of the guided modes in these fibers [24] . This is because the core index is the same as the index of the outer cladding without air holes. Confinement losses can be reduced by increasing the number of rings of air holes which surround the solid core, and thus are determined by the geometry of the structure. It is important to know how many rings of air holes are required to reduce the confinement loss under the Rayleigh scattering limit. At the same time, geometry of the structure allows the chromatic dispersion to be easily controlled by varying the air hole diameter and the pitch of PCF. Here we show that some properties such as losses, ZDW and intermodal dispersion can be controlled keeping the ratio between fiber pitch and hole diameter of the microstructured cladding equal to the called number golden ratio, i.e., 1.618.

Controllability of chromatic dispersion in PCFs is a very important problem for practical applications in optical communication systems, and nonlinear optics. The chromatic dispersion profile can be easily controlled by varying the values of any or all of the following parameters: air hole diameter (d), pitch (Λ), number of rings (Nr) and the number of missing holes (MNr). The aim of this work is to demonstrate how the dispersion profile may be modified by variation of the pitch and the hole diameter, keeping the golden ratio between them, i.e., keeping a ratio between them equal to 1.618, providing a tunable ZWD in the range of 725 nm to 2000 nm.

2. CHARACTERISTICS OF THE FIBER

Figure 1 shows the geometry of the solid core PCF proposed for SCG. The PCF consists of an arrangement of six rings with hexagonal structure and pitch size from 1 to 2 μm (Λ) and air hole diameter from 0.618 to 1.236 μm. This PCF has small effective area and tunable zero wavelength dispersion points in the range from 725 to 2000 nm. Although PCFs can be characterized by variety of methods including finite-difference time-domain method [25] we have used finite-element (FEM) approximation of the Helmholtz equation applying Sellmeier's equation to take the material dispersion into account when calculating effective mode index. The curve of total chromatic dispersion of the PCFs is obtained using following expression

$$D(\lambda) = -\frac{\lambda}{c}\frac{d^2\mathrm{Re}\{n_{\mathrm{eff}}\}}{\mathrm{d}\lambda^2} \tag{2}$$

where c is the speed of light in vacuum, n_{eff} is the effective mode index, and λ is the radiation wave- length.

This structure, proposed to tune the ZDW coinciding with the pump laser wavelengths to facilitate SCG or pulse reshaping applications, is inspired by the nature where the golden ratio can be found in perfect structures of the leaves, animals or the human body. Thus, we propose to use such a golden ratio in the design of a PCF and evaluate its impact on the optical properties. When this proportion is satisfied in the nature, the efficiency of the structures in macro- and micro-world is increased a lot, independently of the eye appeal. Talking about micro-world, we can found the golden ratio in the structure of some viruses, living cells, deoxyribonucleic acid, and bodies of honey bee, leaves or animals. In the macro-world, golden ratio can be found in the structure of galaxies. Due to this we probed to use the golden ratio for synthesis of photonic microstructures just like it happened in architecture to create some design of buildings, for example, keeping this relation. We found that this relation can improve the properties of the PCF; when we keep pitch-to-hole-diameter ratio equal to the golden ratio, it helps us to obtain simpler structures with better characteristics corresponding to the application of need.

In Figure 2 and Figure 3 we show the effective area and nonlinear coefficient of the designed PCF with the pitch Λ = 1, 1.5, 2 μm and air-holes diameter d = 0.618, 0.927, 1.236 μm. Using only six rings of air holes surrounding the solid core we have good confinement of the fundamental mode, small effective area, and high nonlinear coefficient. The structure keeps the golden ratio between pitch and diameter, and with this, we achieve an improvement of the characteristics of PCF correspondingly to an application needed. The effective area and the nonlinear coefficient have been calculated applying the concepts of conventional fiber theory [26] with modifications necessary to include photonic-crystal cladding into account [26] -[28] .

$$V_{\text{eff}} = \frac{2\pi}{\lambda} a_{\text{eff}} \sqrt{n_{co}^2 - n_{\text{FSM}}^2} \tag{3}$$

$$\frac{w}{a_{\text{eff}}} = 0.65 + \frac{1.619}{V_{\text{eff}}^{3/2}} + \frac{2.879}{V_{\text{eff}}^{6}} \tag{4}$$

$$A_{\text{eff}} \cong k_n \pi w^2 \tag{5}$$

$$\gamma(\lambda) = 2\pi n_2 / (\lambda A_{\mathrm{eff}}) \tag{6}$$

where a_{eff} is the effective core radius, λ is the operating wavelength, n_{co} is the core index, n_{FSM} is the effective index of the fundamental space-filling mode, w is the half of mode field diameter (MFD) called also the effective modal spot size, A_{eff} is the effective mode area, k_n is the correction factor [29] , γ is the nonlinear coefficient [30] and n_2 is the nonlinear index of silica. In Section 4 we show an application of the supercontinuum generation using this specific value of nonlinear coefficient shown in Figure 3.

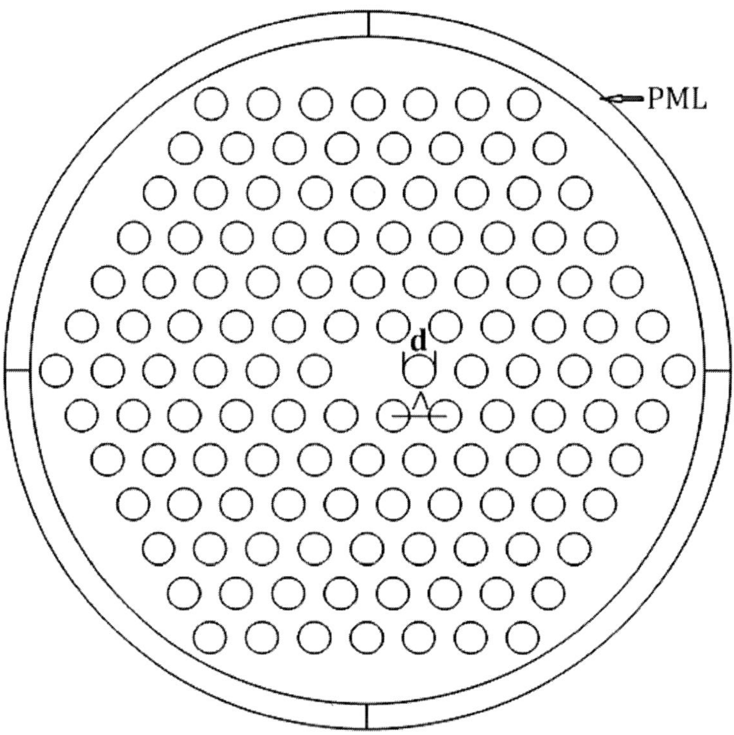

Figure 1. Cross-section of the PCF proposed where we keep the golden ratio between pitch and air hole diameter (Λ/d).

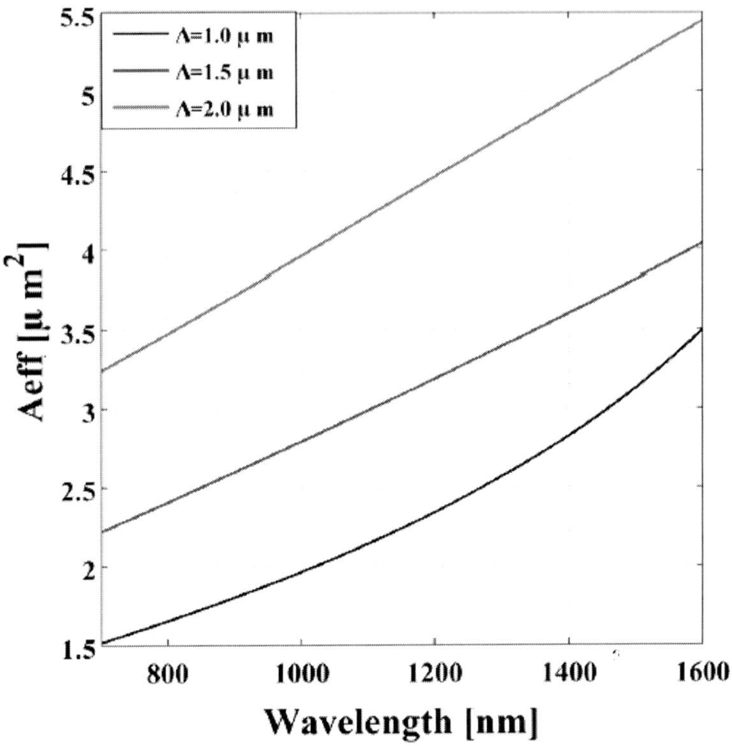

Figure 2. Effective area with Λ = 1, 1.5, 2 μm and d = 0.618, 0.927, 1.236 μm.

3. TUNABLE ZERO DISPERSION WAVELENGTH

In Figure 4 we show the calculated total chromatic dispersion obtained using submicron air hole diameters in order to keep the golden ratio parameter. We can see that varying the pitch from 1 to 1.5 μm two different ZDW points are obtained which are tunable depending on the value of the pitch. Moreover, the curve becomes more gently sloped with increasing the value of pitch, giving the possibility to use this kind of fiber in different applications. In Section 4 we use this design in simulation of super continuum generation. The curves show a maximal chromatic dispersion from

38 to 80 ps/(nm*km) and always we obtain two ZWD points which can be used for pumping by two different lasers to generate a more wide and flattened SC spectrum [31] -[33] .

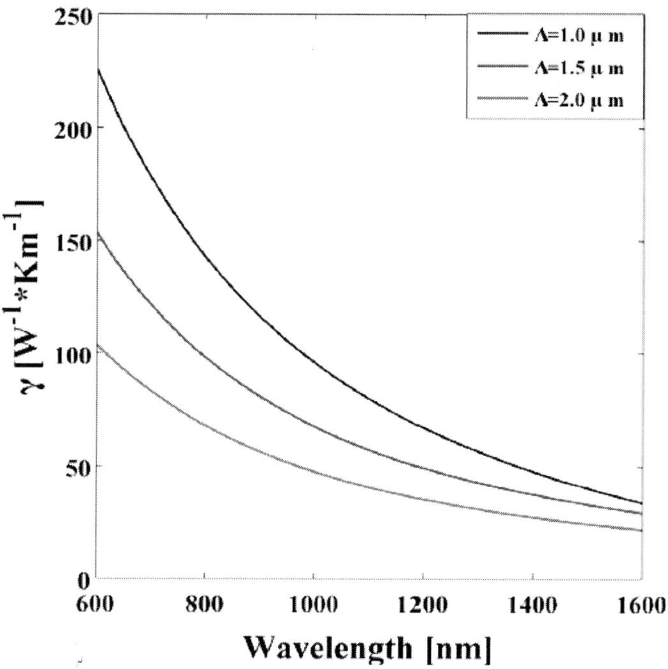

Figure 3. Nonlinear coefficient with Λ = 1, 1.5, 2 μm and d = 0.618, 0.927, 1.236 μm.

In Figure 5 we show chromatic dispersion curves obtained when air hole diameters exceed 1 μm. Therefore, the pitch shall be from 1.618 to 2 μm in order to maintain the golden ratio proportion. The curves shows a maximal chromatic dispersion from 85 to 90 ps/(nm*km), and, as in the previous case, we always have two ZWD points in the range of 725 - 2000 nm. We believe that fabrication of this kind of fiber is simple due to micrometric scale of the constituents.

4. SIMULATION OF SUPER CONTINUUM GENERATION, PULSE RESHAPING AND COMPRESSION

The PCFs proposed here were analyzed using commercial software COMSOL Multiphysics which uses the proven finite element method (FEM). This software runs the finite element analysis together with adaptive meshing and error control using a variety of numerical solvers. The RF Module of COMSOL provides the capabilities to design and is used to simulate electromagnetic wave propagation in PCFs. The mode analysis is made on a cross-section in the xy-plane of the fiber, by expressing the propagating wave in z direction in the following form

$$\boldsymbol{E}(x,y,z,t)=\boldsymbol{E}(x,y)\mathrm{e}^{j(\omega t-\beta z)} \tag{7}$$

and, thereupon, solving the homogeneous Helmholtz equation

$$\nabla\times(\nabla\times\boldsymbol{E})-k_0^2 n^2\boldsymbol{E}=0 \tag{8}$$

for the eigenvalue $\lambda=-j\beta$ applying circular perfectly matched layer, with β being the propagation constant.

The pulse propagation in the PCFs is numerically investigated within the Generalized Nonlinear Schrödinger Equation (GNLSE) [34] [35]

$$\frac{\partial A}{\partial z}=-\frac{\alpha}{2}-\left(\sum_{n\geq 2}\beta_n\frac{i^{n-1}}{n!}\frac{\partial^n}{\partial T^n}\right)A+i\gamma\left(1+\frac{1}{\omega_0}\frac{\partial}{\partial T}\right)\left((1-f_R)A|A|^2+f_R A\int_0^\infty h_R(\tau)|A(z,T-\tau)|^2\,\mathrm{d}\tau\right) \tag{9}$$

where α is the attenuation constant, A is the complex field envelope, z is the distance, τ is the retarded time traveling at the envelope group velocity, γ is the nonlinear coefficient and β_n are the dispersion coefficients obtained by a Taylor series expansion of the propagation constant $\beta(\omega)$ around the center frequency. The GNLSE is solved by Fourth-Order Runge-Kutta in the Interaction Picture Method (RK4IP) [36] [37] .

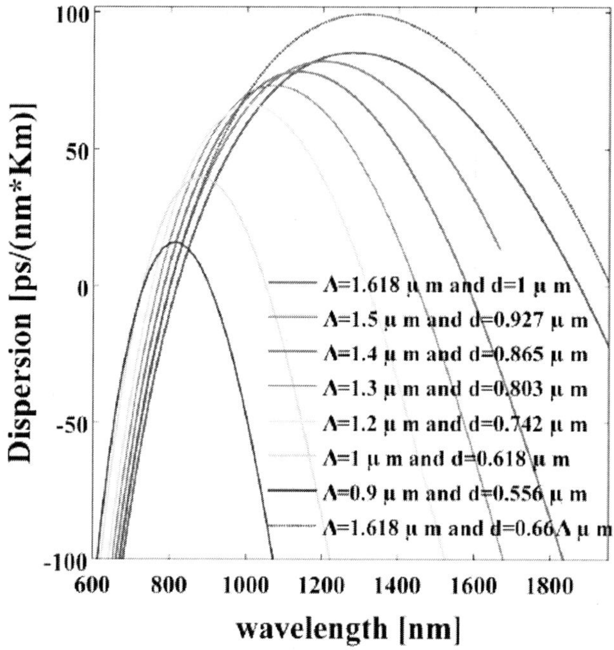

Figure 4. A family of the chromatic dispersion curves calculated by keeping the golden ratio. The ZDW points are moved from 725 to 1800 nm when pitch and air hole diameter are varied.

The Raman response function for silica fibers is described by [38]

$$R(t)=(1-f_r)\delta(t-\tau_e)+f_r h_R(t),\quad h_R(t)=\frac{\tau_1^2+\tau_2^2}{\tau_1\tau_2^2}\exp(-t/\tau_2)\sin(t/\tau_1) \tag{10}$$

To simulate the super continuum generation we consider the sech^2 waveform with the full width at half maximum (FWHM), T_{FWHM} = 50 fs, and 10 kW of peak power and 873 nm, propagating into a 5-mm length of a PCF with maximal chromatic dispersion at 873 nm of 39.78 ps/(nm*km). The PCF designed has low losses, small effective area (A_{eff} = 1.76 μm² at 873 nm), and large nonlinear coefficient in the range of 800 to 1600 nm. Pulse evolution has been simulated using RK4IP method for solution of the generalized nonlinear Schrödinger equation The fiber parameters used in the calculation are given as $\gamma = 122.5\ \mathrm{W^{-1}km^{-1}}$, $f_r = 0.18$, $\lambda_c = 873\ \mathrm{nm}$, $\beta_2 = -16.09\ \mathrm{ps^2/km}$, $\beta_3 = 0.01686\ \mathrm{ps^3/km}$, $\beta_4 = 0.0001732\ \mathrm{ps^4/km}$ in which we achieve a more wide spectrum that reported using similar parameters.

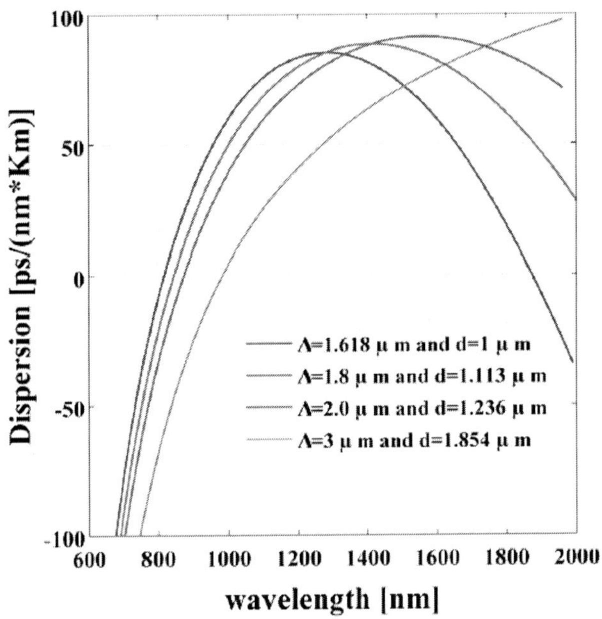

Figure 5. The behavior of chromatic dispersion with variation of geometry parameters, when diameter of holes is greater than 1 μm.

Figure 6 shows the super continuum spectrum obtained by simulation of ultra short pulse propagation in the designed PCF, where we use the property of anomalous dispersion with high nonlinear coefficient of the PCF proposed, for creation a wide spectrum using femtosecond pump laser and shorter length of PCF. We can see a relatively smooth spectrum and wide SC spectrum of

2100 nm bandwidth at the level of −40-dB. In Figure 7 we show temporal evolution of the pump pulse. We observe that soliton-like pump pulse has broken onto seven waveforms which can be interpreted as pulses; it gives the output pulse a form of the train of very short pulses. Taken this observation into account together with the fact that wavelength of the pump pulse is in anomalous dispersion region of the PCF designed, we conclude that Figure 7 show us manifestation of the soliton fission, which is known as a primary nonlinear effect under these propagation conditions [7] . In the spectral domain the soliton fission manifests itself as an extreme broadening of the spectrum of incident pulse. It is exactly what we observe in Figure 6. Thus, the main effect of spectral broadening in our case is the soliton fission.

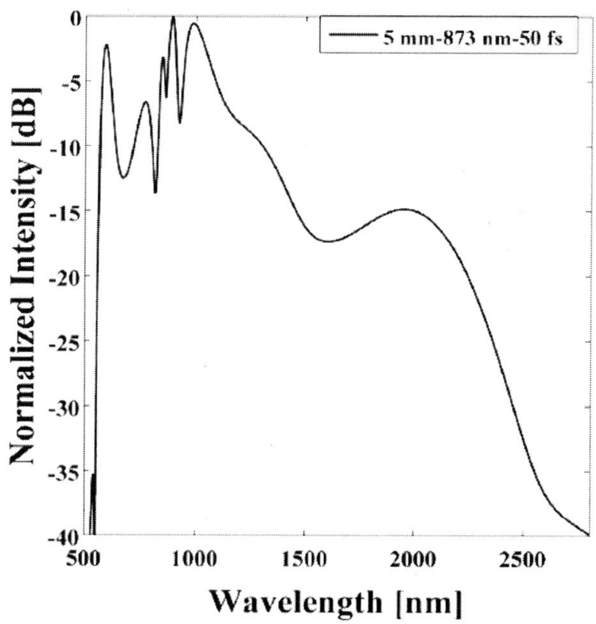

Figure 6. Output spectrum obtained at fiber length of 5 mm. The curve show SC spectrum with the 40-dB bandwidth of 2100 nm using a pump laser centered at 873 nm.

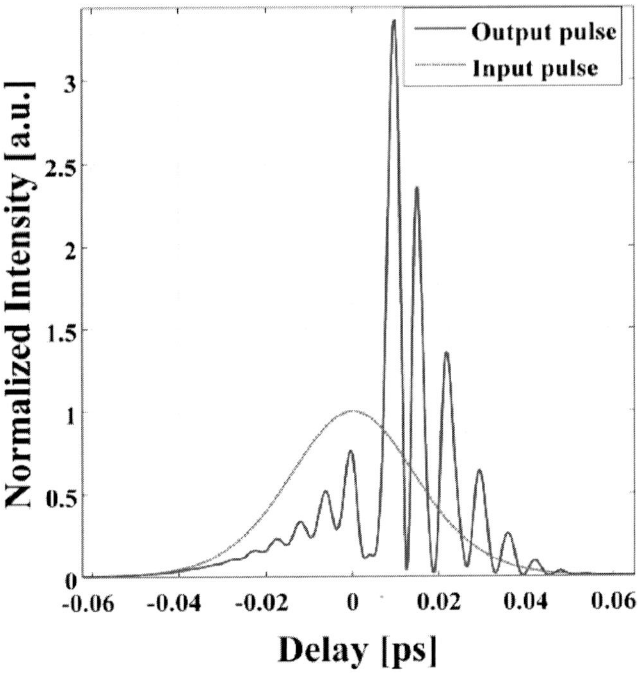

Figure 7. The red curve shows the initial pulse (sech^2 waveform) and the blue curve shows the waveform after propagation in 5 mm-long PCF.

The structure of the PCF designed has also been used for investigation of a possibility of pulse compression and transformation of the pulse shape supposing a sech^2 initial pulse envelope (IPE) waveform. Each of these applications requires a particular combination of dispersion and nonlinearity. Transformation of sech^2 pulse to another high-quality waveform in a fiber requires usually a normal dispersion at the central wavelength of the incident pulse [8] [39] , as the primary phenomenon in this case is an interplay between the self-phase modulation and normal dispersion [8] [40] -[42] . Therefore, for simulation of pulse shape transformation we use following parameters and central wavelength of the pulse to be 600 nm: T_{FWHM} = 140 fs, peak power P = 8 kW, fiber length 13.5 mm, $\gamma = 102.5\ \text{W}^{-1}\text{km}^{-1}$, $f_r = 0.18$, $\beta_2 = 38\ \text{ps}^2/\text{km}$, $\beta_3 = 0.0385\ \text{ps}^3/\text{km}$, $\beta_4 = -0.00002147\ \text{ps}^4/\text{km}$. The wavelength of 600 nm has been chosen because of all dispersion curves we obtained using golden ratio, demonstrate normal dispersion at this wavelength; and this is exactly what we need to achieve pulse transformation with sufficiently high quality of the output pulse.

All of golden-ratio combinations of pitch and hole diameter demonstrate highly nonlinear dispersion profile; therefore, those fibers cannot be applied directly to pulse compression like it made in case of chirped mirrors, for example [43] [44] . However, these combinations provide wavelength ranges with anomalous group velocity dispersion where formation of higher-order solitons is possible. Thus, within these ranges one can compress the pulse using soliton-effect compression [45] . This has been implemented for the case of Λ = 1.618 μm and air-holes diameter $d = (2/3)\Lambda$ with five rings of holes in the cladding. In this case, zero dispersion wavelengths are 800 nm and 1956 nm, and maximum of the chromatic dispersion equals to 99.27 ps/(nm*km) @1303 nm, as shown inFigure 4. Other parameters are following: TFWHM = 250 fs @1303 nm, peak power P = 10 kW, $\gamma = 33\ \text{W}^{-1}\text{km}^{-1}$, $f_r = 0.18$, $\beta_2 = -90\ \text{ps}^2/\text{km}$, $\beta_3 = 0.1262\ \text{ps}^3/\text{km}$, $\beta_4 = -0.0001093\ \text{ps}^4/\text{km}$. The optimal length of the fiber when maximum compression is achieved, equals to 15 mm. The compression factor in this case equals to 100. Figure 8 shows the incident pulses for cases of shape transformation and compression, compressed pulse, and reshaped pulse with a flat-top waveform. The shape of the compressed pulse demonstrates a pedestal, which is typical for pulses obtained via soliton-effect compression. Spectra of corresponding output pulses are shown in Figure 9. Spectrum of the flat-top pulse demonstrates signatures of the self-phase modulation. This nonlinear effect together with normal group velocity dispersion at 600 nm produces flat top pulses. A closer look to the reshaped pulse in Figure 8 allows noting oscillations at the trailing edge of the pulse, which is manifestation of the optical wave breaking [46] .

These results indicate a possibility to use the PCF proposed in variety of engineering problems of nonlinear optics.

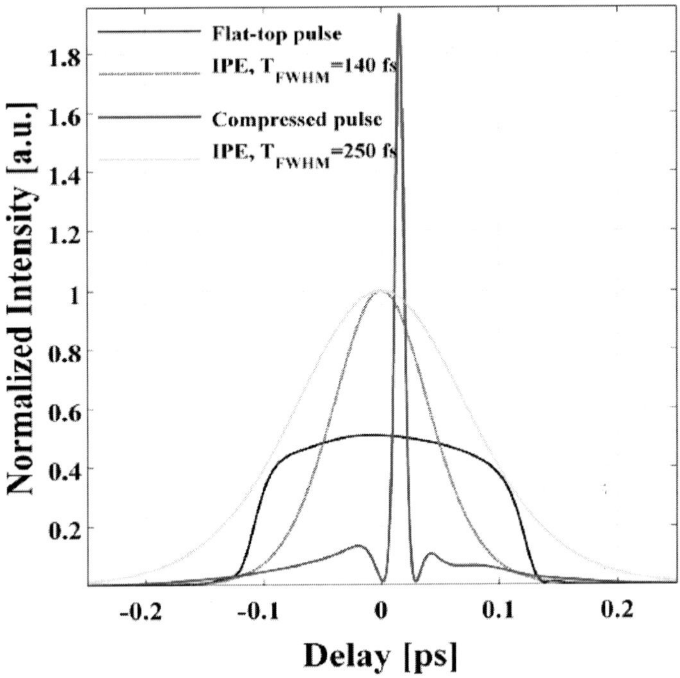

Figure 8. Initial pulse envelope (green and red curve), compressed pulse envelope (blue curve) and flat-top pulse (black curve).

5. CONCLUSION

In this work, we proposed to evaluate how the optical properties of a PCF were influenced when the proportion between pitch and diameter of holes was equal to the golden ratio. In particular, when this ratio is maintained and the pitch and diameter holes are varied, the ZDW could be tunable between the range of 725 nm to 2000 nm. Additionally, with a low number of rings, we obtain a small loss confinement with small effective area and high nonlinear coefficient. Finally, we obtain a flattened chromatic dispersion curve between of two ZDW points. We present an example of applications of this kind of designs for supercontinuum generation, pulse recompression and pulse reshaping. A good quality of flattened and wide spectrum is achieved. The spectrum possesses 2100 nm bandwidth at 40 dB level. The compression factor achieved is 100 producing a compressed pulse of 2.5 fs from an initial pulse of 250 fs. It is believed that the design proposed could be fabricated easily with techniques available nowadays.

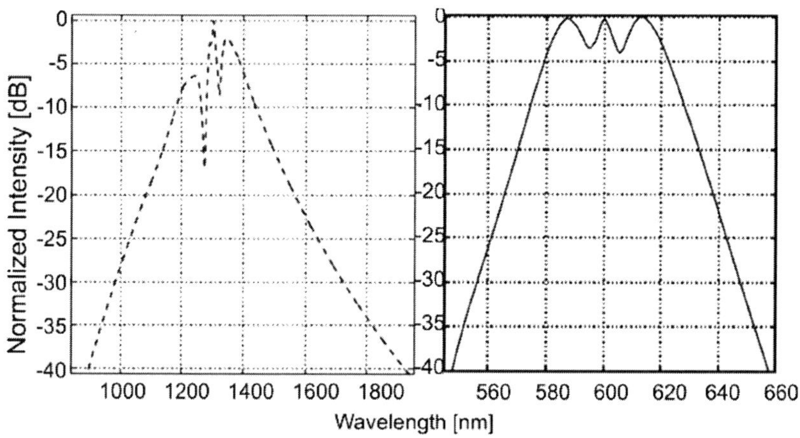

Figure 9. Spectrum for a compressed pulse and flat-top pulse. The dash line show the spectrum of compressed pulse and solid line show the spectrum of the flat-top pulse.

ACKNOWLEDGEMENTS

Alejandro Barrientos García, Igor A. Sukhoivanov, José Amparo Andrade Lucio, Oscar Gerardo Ibarra Manzano, Igor Guryev, Juan Carlos Hernández García, Gabriel Ramos Ortiz (CONACYT) of Mexico, and by Universidad de Guanajuato (project DAIP-192/13 and DAIP-430/2014), FOMIX (CONACyT & CONCyTEG) under grant GTO-2012-

C04-195229 and DAIP 458/2014. A. Barrientos-García and J. Carlos Hernandez-Garcia are grateful to CONACYT for a student research grant No. 329409/336847 and CONACYT project no. 3155.

REFERENCES

1. Knight, J.C. (2003) Photonic Crystal Fibres. Nature, 424, 847-851.

2. Russell, P. (2003) Photonic Crystal Fibers. Science, 299, 358-362.

3. Böswetter, P., Baselt, T., Ebert, F., Basan, F. and Hartmann, P. (2011) Group-Velocity Dispersion in Multimode Photonic Crystal Fibers Measured Using Time-Domain White-Light Interferometry. Proceedings of SPIE, 7914, Article ID: 791422.

4. Liao, M., Chaudhari, C., Qin, G., Yan, X., Suzuki, T. and Ohishi, Y. (2009) Tellurite Microstructure Fibers with Small Hexagonal Core for Supercontinuum Generation. Optics Express, 17, 12174-12182.

5. Gaeta, A.L. (2002) Nonlinear Propagation and Continuum Generation in Microstructured Optical Fibers. Optics Letters, 27, 924-926.

6. Reeves, W.H., Skryabin, D.V., Biancalana, F., Knight, J.C., Russell, P.St.J., Omenetto, F.G., Efimov, A. and Taylor, A.J. (2003) Transformation and Control of Ultra-Short Pulses in Dispersion-Engineered Photonic Crystal Fibres. Nature, 424, 511-515.

7. Dudley, J.M., Genty, G. and Coen, S. (2006) Supercontinuum Generation in Photonic Crystal Fibers. Reviews of Modern Physics, 78, 1135-1184.

8. Sukhoivanov, I.A., Iakushev, S.O., Shulika, O.V., Díez, A. and Andrés, M. (2013) Femtosecond Parabolic Pulse Shaping in

Normally Dispersive Optical Fibers. Optical Express, 21, 17770-17785.

9. Johnson, S.G. and Joannopoulos, J.D. (2001) Block-Iterative Frequency-Domain Methods for Maxwell's Equations in a Planewave Basis. Optics Express, 8, 173-190.

10. White, T.P., Kuhlmey, B.T., McPhedran, R.C., Maystre, D., Renversez, G., de Sterke, C.M. and Botten, L.C. (2002) Multipole Method for Microstructured Optical Fibers. I. Formulation. Journal of the Optical Society of America B, 19, 2322-2330.

11. Laegsgaard, J. and Roberts, P.J. (2008) Dispersive Pulse Compression in Hollow-Core Photonic Band-Gap Fibers. Optics Express, 16, 9628-9644.

12. Lim, H. and Wise, F. (2004) Control of Dispersion in a Femtosecond Ytterbium Laser by Use of Hollow-Core Photonic Band-Gap Fiber. Optics Express, 12, 2231-2235.

13. Dudley, J.M. and Taylor, J.R. (2010) Supercontinuum Generation in Optical Fibres. Cambridge University Press, Cambridge, 32-96.

14. Ferrando, A., Silvestre, E., Miret, J.J., Monsoriu, J.A., Andres, M.V. and Russell, P.St.J. (1999) Designing a Photonic Crystal Fiber with Flattened Chromatic Dispersion. Electronics Letters, 35, 325-326.

15. Ferrando, A., Silvestre, E., Miret, J.J. and Andres, P. (2000) Nearly Zero Ultraflattened Dispersion in Photonic Crystal Fibers. Optics Letters, 25, 790-792.

16. Reeves, W.H., Knight, J.C., Russell, P.St.J. and Roberts, P.J. (2002) Demonstration of Ultra-Flattened Dispersion in Photonic Crystal Fibers. Optics Express, 10, 609-613.

17. Birks, T.A., Mogilevstev, D., Knight, J.C. and Russell, P.St.J. (1999) Dispersion Compression Using Single Material Fibers. IEEE Photonics Technology Letters, 11, 674-676.

18. Knight, J.C., Arriaga, J., Birks, T.A., Ortigosa-Blanch, A., Wadsworth, W.J. and Russell, P.St.J. (2000) Anomalous Dispersion in Photonic Crystal Fiber. IEEE Photonics Technology Letters, 12, 807-809.

19. Mogilevtsev, D., Birks, T.A. and Russell, P.S.J. (1998) Group-Velocity Dispersion in Photonic Crystal Fibers. Optics Letters, 23, 1662-1664.

20. Knight, J.C., Arriaga, J., Birks, T.A., Ortigosa-Blanch, A., Wadsworth, W.J. and Russell, P.S.J. (1998) Anomalous Dispersion in Photonic Crystal Fiber. IEEE Photonics Technology Letters, 12, 817-819.

21. Cheng, H., Crutchfield, W.Y., Doerg, M. and Greengard, L. (2004) Fast, Accurate Integral Equation Methods for the Analysis of Photonic Crystal Fibers I: Theory. Optics Express, 12, 3791-3805.

22. Saitoh, K., Koshiba, M., Hasegawa, T. and Sasaoka, E. (2003) Chromatic Dispersion Control in Photonic Crystal Fibers: Application to Ultra Flattened Dispersion. Optics Express, 11, 843-852.

23. Tajima, K., Zhou, J., Kurokawa, K. and Nakajima, K. (2011) Low Water Peak Photonic Crystal Fibers. European Conference on Optical Communication (ECOC), Rimini, September 2003, Th4.1.6.

24. White, T.P., McPhedran, R.C., de Sterk, C.M., Botten, L.C. and Steel, M.J. (2001) Confinement Losses in Microstructured Optical Fibers. Optics Letters, 26, 1660-1662.

25. Shulika, O., Guryev, I., Gurieva, N., Sukhoivanov, I. and Andrade Lucio, J.A. (2013) Graphics Processing Unit-Acce- lerated Finite-

Difference Time-Domain Method for Characterization of Photonic Crystal Fibers. Optical Engineering, 52, Article ID: 126109.

26. Marcuse, D. (1977) Loss Analysis of Single-Mode Fiber Splices. Bell System Technical Journal, 56, 703-718.

27. Koshiba, M. and Saitoh, K. (2004) Applicability of Classical Optical Fiber Theories to Holey Fibers. Optics Letters, 29, 1739-1741.

28. Mortensen, N.A., Folkenberg, J.R., Nielsen, M.D. and Hansen, K.P. (2003) Modal Cut-Off and the V Parameter in Photonic Crystal Fibers. Optics Letters, 28, 1879-1881.

29. Miyagi, K., Namihira, Y., Razzak, S.M.A., Kaijage, S.F. and Begum, F. (2010) Measurements of Mode Field Diameter and Effective Area of Photonic Crystal Fibers by Far-Field Scanning Technique. Optical Review, 17, 388-392.

30. Liu, Z.L., Hou, L.T. and Wang, W. (2009) Tailoring Nonlinearity and Dispersion of Photonic Crystal Fibers Using Hybrid Cladding. Brazilian Journal of Physics, 39, 50-54.

31. Frosz, M., Falk, P. and Bang, O. (2005) The Role of the Second Zero-Dispersion Wavelength in Generation of Supercontinua and Bright-Bright Soliton-Pairs across the Zero-Dispersion Wavelength. Optics Express, 13, 6181-6192.

32. Hilligsøe, K.M., Andersen, T., Paulsen, H., Nielsen, C., Mølmer, K., Keiding, S., Kristiansen, R., Hansen, K. and Larsen, J. (2004) Supercontinuum Generation in a Photonic Crystal Fiber with Two Zero Dispersion Wavelengths. Optics Express, 12, 1045-1054.

33. Tse, M.L.V., Horak, P., Poletti, F., Broderick, N.G., Price, J.H., Hayes, J.R. and Richardson, D.J. (2006) Supercontinuum Generation at

1.06μm in Holey Fibers with Dispersion Flattened Profiles. Optics Express, 14, 4445-4451.

34. Shulika, O.V., Sukhoivanov, I.A., Guryev, I.V., Andrade-Lucio, J.A., Ibarra-Manzano, O., Barrientos-Garcia, A. and Ramos-Ortiz, G. (2013) Numerical Study of Few-Cycle Pulses Generation from Supercontinuum in ANDi-PCF. Acta Universitaria, 23, 40-44.

35. Shulika, O.V., Sukhoivanov, I.A., Ramos-Ortiz, G., Guryev, I.V., Iakushev, S.O. and Andrade Lucio, J.A. (2013) Characterization of All-Normal Dispersion Microstructured Optical Fiber via Numerical Simulation of Passive Nonlinear Pulse Reshaping and Single-Pulse Flat-Top Supercontinuum. Journal of Nanophotonics, 8, Article ID: 083890.

36. Hult, J. (2007) A Fourth-Order Runge-Kutta in the Interaction Picture Method for Simulating Supercontinuum Generation in Optical Fibers. Journal of Lightwave Technology, 25, 3770-3775.

37. Hernandez-Garcia, J.C., Pottiez, O., Estudillo-Ayala, J.M. and Rojas-Laguna, R. (2012) Numerical Analysis of a Broadband Spectrum Generated in a Standard Fiber by Noise-Like Pulses from a Passively Mode-Locked Fiber Laser. Optics Communications, 285, 1915-1919.

38. Blow, K.J. and Wood, D. (1989) Theoretical Description of Transient Stimulated Raman Scattering in Optical Fibers. IEEE Journal of Quantum Electronics, 25, 2665-2673.

39. Iakushev, S.O., Shulika, O.V. and Sukhoivanov, I.A. (2012) Passive Nonlinear Reshaping towards Parabolic Pulses in the Steady-State Regime in Optical Fibers. Optics Communications, 285, 4493-4499.

40. Agrawal, G. (2012) Nonlinear Fiber Optics. 5th Edition, Academic Press, Waltham, 648 p.

41. Iakushev, S.O., Shulika, O.V., Sukhoivanov, I.A., Fesenko, V.I., Andrés, M.V. and Sayinc, H. (2014) Formation of Ultrashort Triangular Pulses in Optical Fibers. Optics Express, 22, 29119-29134.

42. Sukhoivanov, I.A., Iakushev, S.O., Shulika, O.V., Andrade Lucio, J.A., Díez, A. and Andrés, M. (2014) Supercontinuum Generation at 800 nm in All-Normal Dispersion Photonic Crystal Fiber. Optics Express, 22, 30234-30250.

43. Yakushev, S.O., Shulika, O.V., Petrov, S.I. and Sukhoivanov, I.A. (2008) Chirp Compression with Single Chirped Mirrors and Its Assembly. Microelectronics Journal, 39, 690-695.

44. Iakushev, S.O., Shulika, O.V., Lysak, V.V. and Sukhoivanov, I.A. (2008) Air-Gap Silicon Nitride Chirped Mirror for Few-Cycle Pulse Compression. Optoelectronics and Advanced Materials—Rapid Communications, 2, 686-688.

45. Agrawal, G. (2008) Application of Nonlinear Fiber Optics. 2nd Edition, Academic Press, Waltham.

46. Iakushev, S.O., Shulika, O.V. and Garcia, J.J.R. (2010) Optical Wave Breaking Cancelation and Formation of Quasi- Parabolic Ultrashort Pulses. 10th International Conference on Laser and Fiber-Optical Networks Modeling, Sevastopol, 12-14 September 2010, 176-179.

7

THRESHOLD CHARACTERISTICS ENHANCEMENT OF A SINGLE MODE 1.55 μM INGAASP PHOTONIC CRYSTAL VCSEL FOR OPTICAL COMMUNICATION SYSTEMS

Abbas Majdabadi[1], Saeid Marjani[2*], Masoud Sabaghi[1]

[1]Laser and Optics Research School, Nuclear Science and Technology Research Institute (NSTRL), Tehran, Iran

[2]Department of Electrical Engineering, Ferdowsi University of Mashhad, Mashhad, Iran

ABSTRACT

In the present work, we investigate threshold characteristics of a single mode 1.55 μm InGaAsP vertical cavity surface emitting laser (VCSEL) with two different optical confinement structures. The device employs InGaAsP active region, which is sandwiched between GaAs/AlGaAs and GaAs/AlAs distributed Bragg reflectors (DBRs). The optical confinement introduced by the oxide aperture or a single defect photonic crystal design with holes etched throughout the whole structure, is compared with previous work. Photonic crystal VCSEL shows 30.86% and 57.02% lower threshold current than that of the similar oxide confined VCSEL and previous results, respectively. This paper provides key results of the threshold characteristics, including the threshold current and the threshold power. Results suggest that, the 1.55 μm InGaAsP photonic crystal VCSEL seems to be the most optimal one for light sources in high performance optical communication systems.

KEYWORDS

Threshold Characteristics, Single Mode, VCSEL, Optical Communication Systems

1. INTRODUCTION

In recent years, the vertical cavity surface emitting lasers (VCSELs) have attracted extremely [1] . VCSEL is one of the key light source used in high performance optical communication systems where single mode operation, high output power, high speed modulation and low manufacture cost are necessary [2] . High optical gain in the active area, high temperature, low threshold current and high thermal conductivity in the reflecting mirrors are the main difficulties in developing VCSELs which are used in the field of optical spectroscopy [3] . When all lateral modes except one are suppressed, a fully single mode, VCSEL operation is achieved. Mainly, a single lobe distribution in a desirable lateral direction defined lasing of the fundamental mode. If carriers are confined to the central part of the device, this modal distribution has best overlaps with the gain radial distribution. In real device, the injected current is far from the central part of the device where the fundamental mode has appreciable amplitude. Consequently, it is necessary to funnel the current into the center of the active region. There have been several approaches to address this issue: selective oxidation [4] , proton implantation [5] or structured tunnel junction [6] . On one hand, narrowing the current aperture causes the reduction of the emitted power and on the other hand, broadening of the current aperture favors multimode operation. Hence, it is necessary to additional structuring of VCSELs for their single mode operation with broader current apertures. It can be assured by anti-resonant profile of the refractive index in the distributed Bragg reflectors (DBRs) [7] , surface etching [8] , surface grating [9] and photonic crystals (PhC) [10] . Since PhC has already verified its ability to select very narrow spectrum of allowed frequencies and discriminate all higher modes [11] , it is extremely attractive.

From one side, the PhC improves the wave guiding, which leads to better mode confinement and hence, to higher efficiency of the stimulated recombination. From another side, it destroys the reflectivity of the DBR. The accurate design can balance between these two effects and improve the device efficiency in terms of reduction of the threshold current and increase of lateral mode discrimination.

Recently, we proposed a structure for decreasing the threshold current of VCSELs [12] and showed that proposed structure decreased the threshold current about 76.52% from 2.3 mA to 0.6 mA. There have been a number of reports on the effect of the photonic crystal confined (PhC) on the modal characteristics by Czyszanowski's group [13] -[15] . However, the structures presented are different laser wavelength (1.3 μm), the etching depth and simulation method (the plane-wave admittance method).

In this paper, the impact of the photonic crystal confined and oxide confined (OC) optical confinement schemes on the threshold characteristics of single mode 1.55 μm InGaAsP VCSEL design similar to that reported in [4] is investigated and compared with previous results [12] [14] [15] . The paper is organized as follows: Section 2 briefly describes the theoretical model; Section 3 provides the details of the VCSEL structures; and Section 4 presents the obtained numerical results. Finally, in Section 5, we conclude.

2. THEORY

In modeling VCSEL, we must the electrical, optical and thermal interaction during VCSEL performance [16] . Thus base of simulation is to solve Poisson and continuity equations for electrons and holes. Poisson's equation is defined by [17] :

$$\nabla\cdot(\varepsilon\nabla\Psi)=\rho \tag{1}$$

where ψ is electrostatic potential, ρ is local charge density and ε is local permittivity. The continuity equations of electron and hole are given by [17] :

$$\frac{\partial n}{\partial t}=G_n-R_n+\frac{1}{q}\nabla\cdot\overrightarrow{J_n}, \tag{2}$$

$$\frac{\partial p}{\partial t}=G_p-R_p+\frac{1}{q}\nabla\cdot\overrightarrow{J_p}. \tag{3}$$

where n and p are the electron and hole concentration, J_n and J_p are the electron and hole current densities, G_n and G_p are the generation rates for electrons and holes, R_n and R_p are the recombination rates and q is the magnitude of electron charge.

The fundamental semiconductor Equations (1)-(3) are solved self-consistently together with Helmholtz and the photon rate equations. The applied technique for solution of Helmholtz equation is based on effective frequency method [18] which shows accuracy for great portion of preliminary problems. Two-dimensional Helmholtz equation is solved to determine the transverse optical field profile and it is given by [17] :

$$\nabla^2 E(r,z,\varphi)+\frac{\omega_0}{c^2}\varepsilon(r,z,\varphi,\omega)E(r,z,\varphi)=0 \tag{4}$$

where ω is the frequency, $\varepsilon(r,z,\varphi,\omega)$ is the complex dielectric permittivity, $E(r,z,\varphi)$ is the optical electric field and c is the speed of light in vacuum.

The light power equation relates electrical and optical models. The photon rate equation is given by [17] :

$$\frac{dS_m}{dt} = \left(\frac{c}{N_{\text{eff}}} G_m - \frac{1}{\tau_{\text{phm}}} - \frac{cL}{N_{\text{eff}}} \right) S_m + R_{\text{spm}} \tag{5}$$

where S_m is the photon number, G_m is the modal gain, R_{spm} is the modal spontaneous emission rate, L represents the losses in the laser, N_{eff} is the group effective refractive index, τ_{phm} is the modal photon lifetime and c is the speed of light in vacuum. The m refers to the modal number.

The heat flow equation has the form [17] :

$$C \frac{\partial T_L}{\partial t} = \nabla \left(\kappa \nabla T_L \right) + H \tag{6}$$

where C is the heat capacitance per unit volume, κ is the thermal conductivity, H is the generation, T_L is the local lattice temperature and H is the heat generation term.

The heat generation equation has the form [17] :

$$H = \left[\frac{\left|\overrightarrow{J_n}\right|^2}{q\mu_n n} + \frac{\left|\overrightarrow{J_p}\right|^2}{q\mu_p p} \right] + q(R-G)\left[\phi_p - \phi_n + T_L \left(P_p - P_n \right) \right] - T_L \left(\overrightarrow{J_n} \nabla P_n + \overrightarrow{J_p} \nabla P_p \right) \tag{7}$$

where: $\left[\frac{\overline{|J_n|^2}}{q\mu_n n}+\frac{\overline{|J_p|^2}}{q\mu_p p}\right]$is the Joule heating term; $q(R-G)\left[\phi_p-\phi_n+T_L\left(P_p-P_n\right)\right]-T_L\left(\overline{J_n}\nabla P_n+\overline{J_p}\nabla P_p\right)$is the recombination and generation heating and cooling term; and the Peltier and Thomson effects.

In addition, all important, usually non-linear, interactions between the optical, electrical, thermal and recombination phenomena are also taken into account with the aid of the self-consistent iteration algorithm, including: gain-induced wave-guiding, thermal focusing, self-focusing, temperature dependence of optical gain and absorption coefficients and the energy gaps and electrical conductivities and thermal conductivities, carrier-con- centration dependence of electrical conductivities and optical gain and absorption coefficients and the energy gaps and wavelength dependences of optical gain and absorption coefficients. Consequently, several factors are effective in determining the profiles of all model parameters within the whole VCSEL volume, including: different chemical compositions of the VCSEL layers, the distributions of the temperature, the current density, the carrier concentration and the mode radiation intensity.

Heat loss from the modeled VCSEL device was specified using thermal contacts at the top electrode, bottom electrode and the device sidewall. The thermal contacts define thermal conductivities to simulate heat loss from radiation via exposed surfaces or conduction through the semiconducting material to a heatsink.

Equation (1)-(7) provide an approach that can account for the mutual dependence of electrical, thermal, optical and elements of heat sources. In this paper, we employ numerical-based simulation software to assist in the device design and simulation [17] .

3. VCSEL STRUCTURE

The analyzed structure is similar to the one recently reported in [12] has been chosen as a model structure for the analysis of the 1.55 μm InGaAsP VCSEL. The active region consists of six quantum wells where the well is 5.5 nm $In_{0.76}Ga_{0.24}As_{0.82}P_{0.18}$ and the barrier is 8 nm $In_{0.48}Ga_{0.52}As_{0.82}P_{0.18}$. In both sides of this active region, there is InP and on top of it, GaAs. The top mirror is 30 layers of GaAs/$Al_{0.33}Ga_{0.67}As$ with index of refraction 3.38 and 3.05 respectively, and the bottom mirror has 28 layers of GaAs/AlAs with index of refraction 3.38 and 2.89 respectively. Layer thickness, doping, majority carrier mobility, refractive index, temperature coefficient of n, absorption coefficient, and thermal conductivity of layers are listed in Table 1. In the OC VCSEL, the incorporation of a high aluminum content layer ($Al_{0.98}Ga_{0.02}As$) in two DBR periods above the active region allows for selective oxidation [11] . Figure 1(a) shows the PhC VCSEL structure, which is similar to the above VCSEL [12] but made additionally with a single defect PhC etched throughout the whole structure. The optical confinement is achieved by means of seven air holes where the center is missed off to make the defect region, as shown in Figure 1(b). The crucial PhC parameters are connected by the air hole diameter (a), the pitch (b) and the optical aperture diameter (r), are defined in Figure 1(b). The transverse index guiding around the single defect region can be controlled by the air hole diameter the pitch ratio (a/b). We fixed the optical aperture diameter at 2 μm and a/b ratio is varied from 0 to 0.95 for the study.

Table 1. Parameters of the structure (l: layer thickness; N_{dop}: doping; u: majority carrier mobility;n: refractive index; dn/dt: temperature coefficient of n; α: absorption coefficient; and κ: thermal conductivity).

Parameter unit	1 (μm)	N_{dop} (1/cm^3)	μ (cm^2/V·s)	n	dn/dT (10^{-4}/K)	α (1/cm)	κ (W/cm·K)
Au/Ti (contact)	0.200	-	-	0.83	-	684,000	0.67
p-GaAs	0.020	2×10^{19}	-	3.38	3	500	0.44
p-GaAs	0.182	4×10^{17}	-	3.38	3	25	0.22
p-$Al_{0.67}Ga_{0.33}As$ (DBR)	0.127	4×10^{17}	-	3.05	2	25	0.22
p-GaAs (DBR)	0.115	4×10^{17}	-	3.38	3	25	0.44
p-GaAs (spacer)	0.020	4×10^{17}	-	3.38	3	25	0.44
p-GaAs (spacer)	0.010	4×10^{19}	-	3.38	3	1000	0.44
p-InP (spacer)	0.178	1×10^{18}	30	3.17	2	24	0.68
p-Inp (spacer)	0.100	1×10^{16}	150	3.17	2	0.24	0.68
$In_{0.76}Ga_{0.24}As_{0.82}P_{0.18}$ (QW)	0.0055	-	100	3.6	2	54	0.043
$In_{0.48}Ga_{0.52}As_{0.82}P_{0.18}$ (barrier)	0.008	-	100	3.4	2	54	0.043
n-InP (spacer)	0.258	5×10^{18}	4600	3.15	2	8	0.68
n-GaAs (spacer)	0.050	1×10^{18}	-	3.38	3	6	0.44
n-GaAs (DBR)	0.115	1×10^{18}	-	3.38	3	6	0.22
n-AlAs (DBR)	0.134	1×10^{18}	-	2.89	1	3	0.22
n-GaAs (substrate)	450	5×10^{18}	-	3.38	3	5.8	0.44

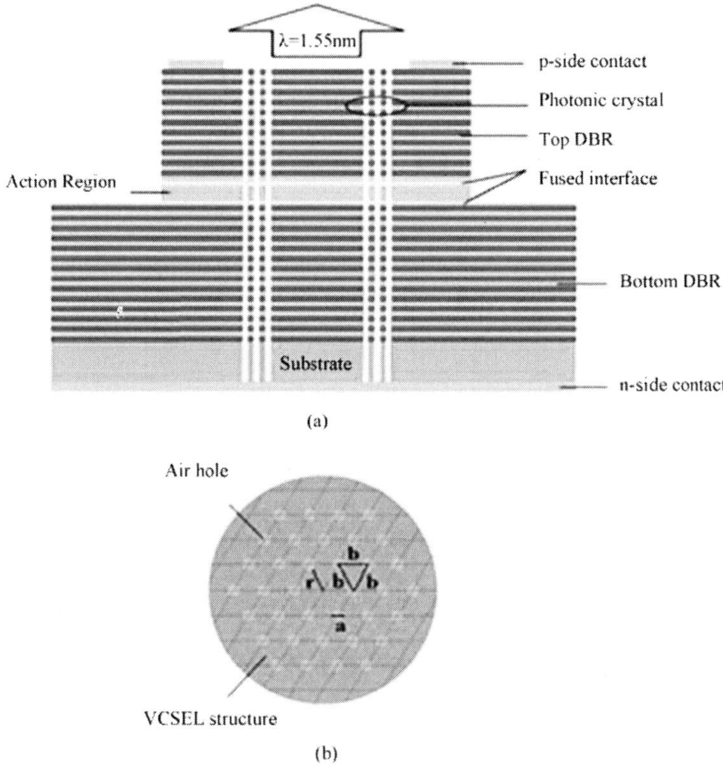

Figure 1. (a) Schematic structure of the VCSEL device and (b) top view of the triangular-lattice air holes pattern.

4. RESULTS

The present work allows for determination of the optimum VCSEL structure that is found by a minimization of the threshold power and current. Figure 2 presents a comparison of the optical intensity distributions at the lasing threshold of OC and PhC VCSELs with different a/b values that emphasize the overlap of the holes by the modes for the cutoff condition. Fundamentally, the increase of a/b ratio and the optical aperture diameter leads to an increase of the optical confinement factor within the active region and the optical intensity distributions changes significantly. The case of a/b = 0.3 shows a star-like shape of the PhC mode that the field permeates the absorptive surrounding regions. However, at higher values of a/b ratios the mode distributions remain almost unchanged.

A comparison of the threshold characteristics of OC, PhC VCSELs and previous work [12] [14] [15] as a function of the a/b ratio is presented in Figures 3-5. As can be seen from Figure 3, the logarithmically scaled threshold power of PhC VCSELs decrease gradually with the increase in the a/b ratio. This initial decrement (from a/b = 0 to a/b = 0.9) should be mainly due to improving the optical confinement by the photonic crystal. For a/b = 0.9 the threshold power reaches a minimum value of 0.5719e−5 W and then rapidly increases for a/b = 0.95, which should be mainly due to blocking of most of the current flow by the holes. The minimal threshold power is lowered by 48.82% from 1.1176e−5 W to 0.5719e−5 W than that of the OC VCSEL, since, the electrical resistance of the etched PhC regions is increased.

Figure 4 shows a monotonic decrease in the logarithmically scaled threshold current of PhC VCSELs with the increase in the a/b ratio. This is caused by more effective optical confinement and stronger limitation of the current flow to the active region. As shown in Figure 4, photonic crystal VCSEL shows 30.86% and 57.02% lower threshold current than that of the similar oxide confined VCSEL and previous results [19] , respectively. Threshold current for two PhC VCSEL structures [14] [15] determined by r = 4 μm, a/b = 0.5 and r = 8 μm, a/b = 0.5 are 4.9 mA and 7.11 mA, respectively, which are very close to previous values of 4.89 mA and 7.12 mA, respectively [15] . We fixed the depth of the PhC holes at 7 μm for two PhC VCSEL structures.

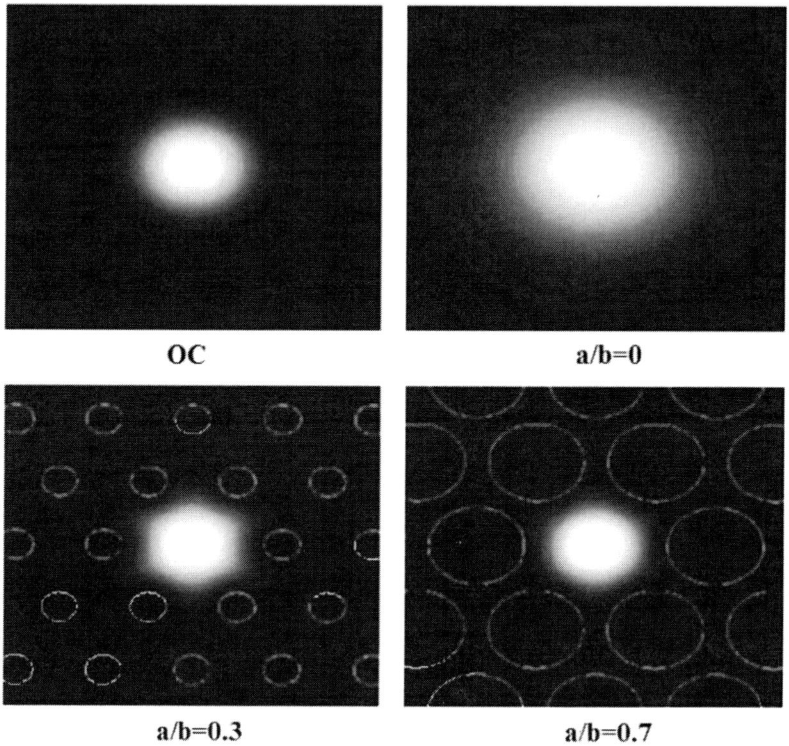

Figure 2. The optical intensity distributions at the lasing threshold of the fundamental mode within the active region in the case of the OC VCSEL and the PhC VCSEL for four different a/b ratios.

Figure 5 shows a threshold temperature increase for a/b = 0 than that of the OC VCSEL, since, the thermal focusing is the unique wave guiding mechanism induced by the PhC structure. As can be seen from Figure 5, threshold temperature of the OC VCSEL is always smaller than that of the PhC VCSEL. This is caused by the more efficient heat sinking process in the OC VCSEL. For a/b = 0.7 the threshold temperature reaches a minimum value of 300.147 K that is the optimum configuration for over threshold operation.

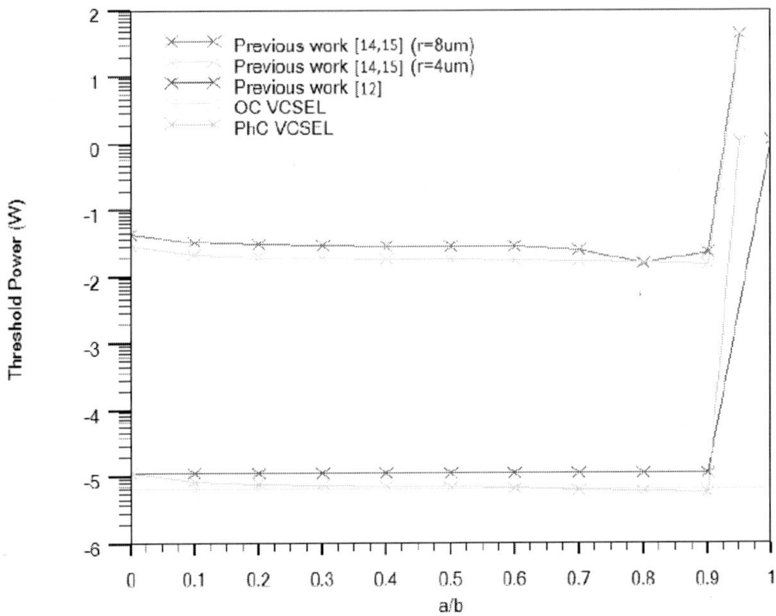

Figure 3. Threshold power of OC, PhC VCSELs and previous work [12] [14] [15] as a function of the a/b ratio.

5. CONCLUSION

Threshold characteristics of a single mode 1.55 μm InGaAsP vertical cavity surface emitting laser with two different optical confinement structures were analyzed and compared with previous work. In summary, the results indicate that lowest threshold current and threshold power is for a/b = 0.9 that shows 30.86% and 57.02%

Abbas Majdabadi, Saeid Marjani, Masoud Sabaghi lower threshold current than that of the similar oxide confined VCSEL and previous results, respectively. Our studies reveal that high-precision PhC etching is necessary in PhC VCSELs in order to optimize their threshold characteristics for light sources in high performance optical communication systems.

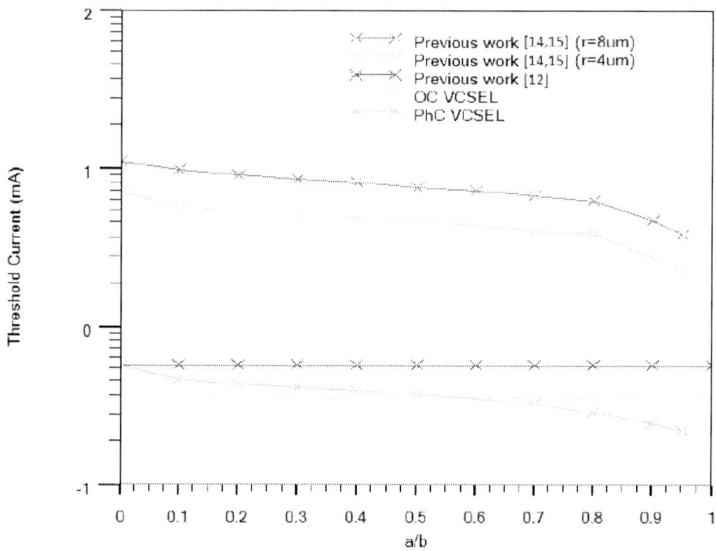

Figure 4. Threshold current of OC, PhC VCSELs and previous work [12] [14] [15] as a function of the a/b ratio.

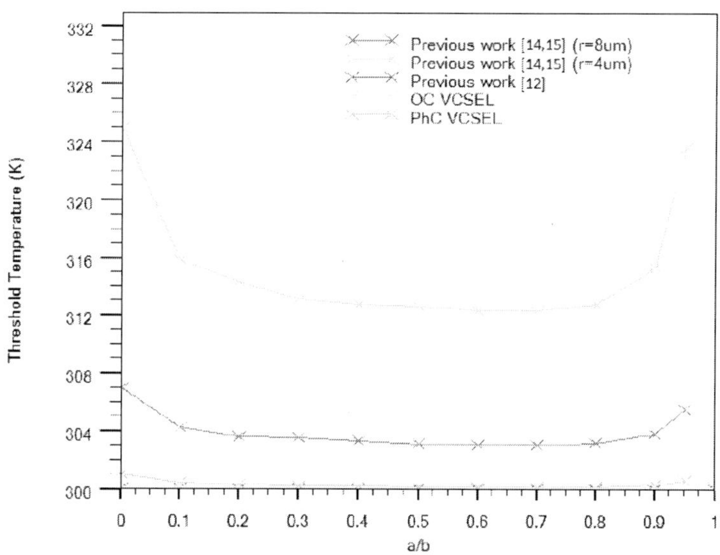

Figure 5. Threshold temperature of OC, PhC VCSELs and previous work [12] [14] [15] as a function of the a/b ratio.

REFERENCES

1. Iga, M.K. (2000) Surface-Emitting Laser—Its Birth and Generation of New Optoelectronic Field. IEEE Journal of Selected Topics in Quantum Electronics, 6, 1201-1215.

2. Dems, M., Kotynski, R. and Panajotov, K. (2005) Plane Wave Admittance Method—A Novel Approach for Determining the Electromagnetic Modes in Photonic Structures. Optics Express, 13, 3196-3207.

3. Kapon, E. and Sirbu, A. (2009) Long-Wavelength VCSELS Power-Efficient Answer. Nature Photonics, 3, 27-29.

4. Jung, C., Jager, R., Grabherr, M., Schnitzer, P., Michalzik, R., Weigl, B., Muller, S. and Ebeling, K.J. (1997) 4.8 mW Single-Mode Oxide Confined Topsurface Emitting Vertical-Cavity Laser Diodes. Electronics Letters, 33, 1790-1791.

5. Morgan, R.A., Guth, G.D., Focht, M.W., Asom, M.T., Kojima, K., Rogers, L.E. and Callis, S.E. (1993) Transverse Mode Control of Vertical-Cavity Top-Surface-Emitting Lasers. IEEE Photonics Technology Letters, 4, 374-377.

6. Long, C.M., Mutter, L., Dwir, B., Mereuta, A., Caliman, A., Sirbu, A., Iakovlev, V. and Kapon, E. (2014) Optical Injection Locking of Transverse Modes in 1.3-μm Wavelength Coupled-VCSEL Arrays. Optics Express, 22, 21137- 21144.

7. Zhou, D. and Mawst, L.J. (2002) High-Power Single-Mode Antiresonant Reflecting Optical Waveguide-Type Vertical- Cavity Surface-Emitting Lasers. IEEE Journal of Quantum Electronics, 38, 1599-1606.

8. Debernardi, P., Unold, H.J., Maehnss, J., Michalzik, R., Bava, G.P. and Ebeling, K.J. (2003) Single-Mode, Single-Po- larization VCSELs via Elliptical Surface Etching: Experiments and Theory.

IEEE Journal of Selected Topics in Quan- tum Electronics, 9, 1394-1404.

9. Haglund, A., Gustavsson, J.S., Bengtsson, J., Jedrasik, P. and Larsson, A. (2006) Design and Evaluation of Fundamental-Mode and Polarization-Stabilized VCSELs with a Subwavelength Surface Grating. IEEE Journal of Quantum Electronics, 42, 231-240.

10. Siriani, D.F., Tan, M.P., Kasten, A.M., Lehman Harren, A.C., Leisher, P.O., Sulkin, J.D., Raftery Jr., J.J., Danner, A.J., Giannopoulos, A.V. and Choquette, K.D. (2009) Mode Control in Photonic Crystal Vertical-Cavity Surface-Emitting Lasers and Coherent Arrays. IEEE Journal of Selected Topics in Quantum Electronics, 15, 909-917.

11. De La Rue, R. (2006) Photonic Crystal Components: Harnessing the Power of the Photon. Optics and Photonics News, 17, 30-35.

12. Faez, R., Marjani, A. and Marjani, S. (2011) Design and Simulation of a High Power Single Mode 1550 nm InGaAsP VCSELs. IEICE Electronics Express, 8, 1096-1101.

13. Czyszanowski, T., Dems, M., Sarzala, R., Nakwaski, W. and Panajotov, K. (2011) Precise Lateral Mode Control in Photonic Crystal Vertical-Cavity Surface-Emitting Lasers. IEEE Journal of Quantum Electronics, 99, 1291-1296.

14. Czyszanowski, T., Dems, M., Thienpont, H. and Panajotov, K. (2008) Modal Gain and Confinement Factors in Top- and Bottom-Emitting Photonic-Crystal VCSEL. Journal of Physics D: Applied Physics, 41, Article ID: 085102.

15. Czyszanowski, T. (2009) Discrimination of Higher-Order Modes in Photonic-Crystal VCSEL. Proceedings of the IEEE/LEOS Winter Topicals Meeting Series, Innsbruck, 12-14 January 2009, 20-21.

16. Menon, P.S., Kumarajah, K., Ismail, M., Majlis, B.Y.M. and Shaari, S. (2010) Long-Wavelength MQW Vertical-Cav- ity Surface Emitting Laser: Effects of Lattice Temperature. Journal of Optical Communications, 31, 81-84.

17. SILVACO International Incorporated (2010) ATLAS User's Manual. Version 5.12.0.R., SILVACO, Inc., USA.

18. Wenzel, H. and Wunsche, H.-J. (1997) The Effective Frequency Method in the Analysis of Vertical-Cavity Surface- Emitting Lasers. IEEE Journal of Quantum Electronics, 33, 1156-1162.

19. Choquette, K.D., Geib, K.M., Ashby, C.I., Twesten, R.D., Blum, O., Hou, H.Q., Follstaedt, D.M., Hammons, B.E., Mathes, D. and Hull, R. (1997) Advances in Selective Wet Oxidation of AlGaAs Alloys. IEEE Journal of Selected Topics in Quantum Electronics, 3, 916-926.

8

ON THE VALIDITY OF THE EFFECTIVE INDEX METHOD FOR LONG PERIOD GRATING PHOTONIC CRYSTAL FIBERS

Dhia Khadri[1,2], Walid Belhadj[3*], Douha Gamra[2], Fathi AbdelMalek[4], Habib Bouchriha[2]

[1]Ecole Polytechnique de Montreal, Genie Physique, Montreal, Canada

[2]Laboratoire de Physique Quantique et Photonique, Faculte des Sciences de Tunis, Tunis, Tunisia

[3]Faculté des Sciences de Bizerte, Bizerte, Tunisia

[4]Department of Physics, Jazan University, Jazan, KSA.

ABSTRACT

This paper focuses on the investigation of modal characteristics and sensing properties of long period grating photonic crystal fibers (LPG-PCFs). An improved effective index method is employed with an objective to study its limitations for various designs of LPG-PCFs. Results so obtained with the above method are compared with the corresponding values of multiple multipole (MMP) method results which points the range of validity and applicability of the improved effective index method to LPG-PCFs. It is shown that this method is excellent when the surrounding media is assumed to be air.

However, it becomes less accurate when the fiber is immersed into a liquid with a refractive index close to that of the cladding.

1. INTRODUCTION

The development of long-period grating (LPG) has had a significant impact on fiber optic sensing. It is known that the spectral response of LPG is sensitive to local environment such as temperature, strain, bend radius and refractive index of the medium surrounding the fiber. LPG sensor is distinct from other fiber optic sensing techniques such as fiber-laser-based sensing [1] as LPG can perform simultaneous multiple parameter measurement. Moreover, the sensitivity of LPG to the surrounding refractive index has been extensively studied to extend the application range, namely in biomedical and chemical sensing [2]. Applications of LPGs in temperature [3], strain [4], and bend [5,6] sensors have since been demonstrated. However, for chemical sensors, it is mainly the sensitivity of LPGs to the surrounding refractive index that is of interest for sensor development. Examples of chemical sensors for sugar [7], aromatic compounds [8], and hydrocarbons [9] based on the detection of refractive index change using LPGs have been presented in the literature. In these applications, direct sensing was used to detect changes in the refractive index of mixtures in a region close to the refractive index of the fiber cladding, which is the most sensitive interval. Indirect sensing of analytes is also possible using a chemically selective layer, again with a refractive index close to that of the cladding [10].

Because of the wide range of applications mentioned above, there is a growing interest in optical refractive index sensors. Those based on photonic crystal fibers (PCFs) or fiber gratings are probably the most attractive ones since they offer high sensitivity, stable wavelengthmodulated information, and broad measuring range. At a refractive index of 1.33 typical for aqueous environments, the sensitivity of a fiber grating based sensor is typically 50 nm/refractive index units (RIU) [11], while PCFbased sensors possess higher sensitivity of more than 1500 nm/RIU [12,13]. The characteristics of LPG made in a PCF strongly depend on its structural parameters such as core radius and refractive indices of the core and cladding [14-17]. Therefore, for design of an LPG based on a PCF, the optimization of PCF parameters is a necessary condition [17].

To determine optimal values of the aforesaid PCF parameters, numerical methods such as the finite element method [18-20], the finite-difference method [21], and the MMP [22] are generally utilised. However, these numerical methods of analyzing PCFs require a great deal of computational time and a large numerical resource. A very simple technique referred to as improved effective index method (IEIM) was reported in [23,24]. It is has been shown that this method enables a full access to study the resonance shifts of the LPG as a function of external refractive index caused by both the first cladding mode and the core mode of the photonic crystal fibre. In this paper we focus on the IEIM described in [23,24]. We make a quantitative investigation of this technique and we discuss their limitations for various designs of LPG-PCFs by comparing its results to those obtained by multipole method (MMP) implemented by Cudos Mof [22].

2. MODELING OF LPG IN PHOTONIC CRYSTAL FIBER BY IEIM

The wavelength response of LPG can be quantitatively analyzed by a coupled mode theory which includes respectively the core mode coupled to the resonant cladding mode, and satisfying the phase matching condition given by [25]:

$$\lambda = \left(n_{eff}^{co} - n_{eff}^{cl}\right)\Lambda_{res} \tag{1}$$

where n_{eff}^{co} and n_{eff}^{cl} are the effective mode indices of the core and cladding modes, respectively, Λ_{res} is the grating period, and λ is the resonant wavelength.

The IEIM applied on PCF offers the possibility to calculate the refractive effective index of core mode [23] and cladding mode [24]. This method proposes the analogy between the analysed PCF and W-shaped fibre with a suppressed cladding. The W-fibre corresponding to PCF is obtained by replacing the holey region with a homogeneous material with dielectric permittivity equal to the averaged permittivity of the air holes and silica. After fitting to the radius of the core, good agreement in propagation constants, mode shapes and the order of the core mode between the two models was achieved.

Extensive research has been carried out to investigate the dependence of resonance wavelength as a function of the refractive index change of the surrounding media [26-30] and a typical calibration curve (**Figure 1**) shows the resonance wavelength change as a function of the refractive index (RI) of the surrounding medium with the cladding refractive index of the LPG being ~1.45. This indicates that when the surrounding refractive index is lower than that of the fibre cladding, the resonance wavelength of the LPG changes with the variation of surrounding refractive index. However, when the surrounding refractive index is higher than that of the fibre cladding, the resonance conditions become more complex. In the case where the surrounding refractive index is higher than that of the cladding, two different situations are considered. The first case is where there is a thick overlay placed on top of the LPG and the second is where this overlay on the LPG is relatively thin. When the LPG is surrounded with a thick overlay that has higher refractive index value than that of the cladding, the fibre does not support any cladding mode and the core mode couples to the radiation modes [31]. The confined field thus forms leaky modes. The resonance losses of the LPG increase with the increase in the external refractive index value as the leaky modes are better confined. The wavelength of the resonance band does not change with the external refractive index value as the phase of the reflected field at the cladding/surrounding interface does not change [31]. In this work we discuss the validity of IEIM method when the surrounding refractive index become close to that of the cladding.

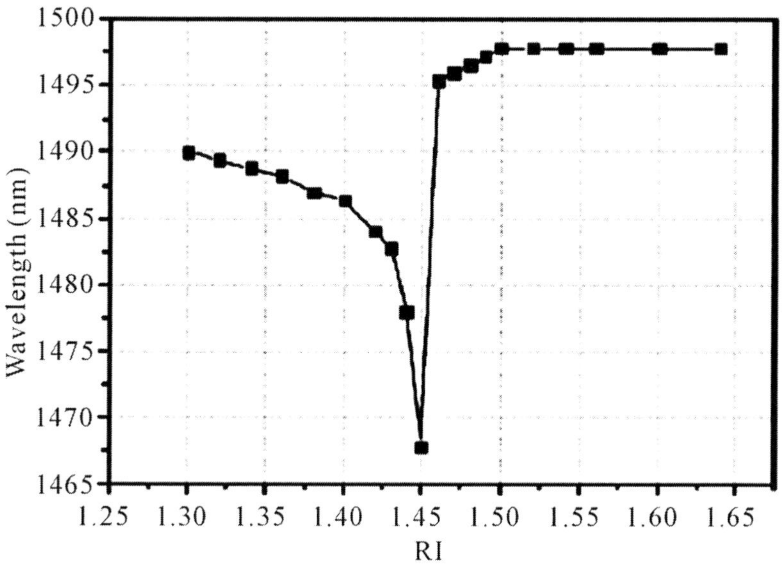

Figure 1. Calibration curve of a typical LPG resonance wavelength as function of the surrounding refractive index.

The structured of the LPG refractive index sensor under study is defined in **Figure 2**. It consists of a cladding which is assumed to be a two-dimensional photonic crystal made of silica (n_{SiO2} = 1.45) with air holes running along the length of the fibre. The holes of diameter d are arranged in a triangular pattern across the cross section with a lattice constant is Λ. The hole pitch and diameter of the considered structure are respectively Λ = 8.0 μm and d = 3.68 μm [32].

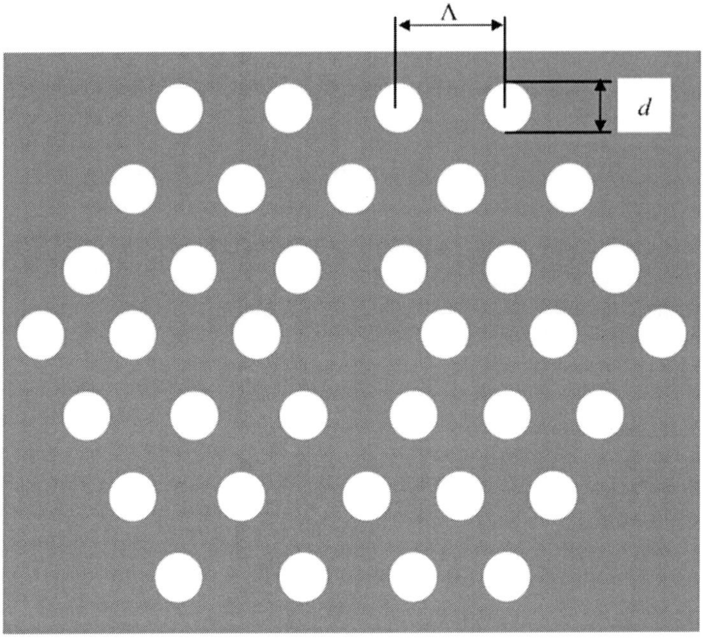

Figure 2. Schematic diagram of the LPG refractive index sensor.

In **Figure 3**, we report the dispersion characteristics of the core mode and the space filling mode for different hole to pitch and wavelength to pitch ratios. It can be seen from this figure that the effective indices decreases by increasing the hole size (d/Λ) or increasing the normalized wavelength (λ/Λ). The effective index of the core mode is greater than that of the space filling mode, regardless to the fibre parameters (d, Λ). The curves in **Figure 4** are obtained by applying the equation (1) and the negative slope of grating characteristics justifies the observed blue shift of the grating spectrum with increasing grating period.

Furthermore, if we limit the analysis to gratings that consist of a circularly symmetric index perturbation in any transverse plane of the fibre, the only nonzero coupling coefficients between the core mode and the cladding modes involve cladding modes of azimuthal order l = 1. Consequently the fundamental core mode LP01, which has ν = 1, can be coupled to the HE_{1m} modes. The Coupling coefficient for two modes is determined by the overlap integral weighted by the transverse refractive index change in the grating region.

$$\kappa_i = \frac{\pi}{\lambda} \int E_{co}(x,y) E^{*}_{cl,i}(x,y) \Delta n(x,y) \mathrm{d}x\mathrm{d}y \tag{2}$$

where Δn presents the dielectric function variation induced by the perturbation.

Due to the fact that the effective refractive indices of both the core and the cladding are dependent on the wavelength at which they are evaluated, any change in ambient refractive index that leads to a wavelength shift for that particular resonant band causes variations in the modal indices as follows: n_{eff}^{co} changes according to the new resonant wavelength, and n_{eff}^{cl} is affected not only by the new ambient index but also by the new resonant wavelength that results. This means that the true spectral shift in wavelength can be calculated from the phasematching condition only if the effective refractive indices have been evaluated at the new resonant wavelength that is reached after the spectral shift has occurred as a result of the change in n_{ext}. The sensitivity of the LPG was calculated by the evaluation of the effective propagation constant of cladding mode versus the external refractive index. Once the value of ambient refractive index reaches that of the cladding, discrete cladding modes are no longer guided along the fibre. This is evident in the absence of distinct attenuation bands in the LPG transmission spectrum at this point (the spectral properties are "washed out") and broadband losses are visible due to lousy coupling with the continuum of radiation modes [33]. However when the index is greater than the index of silica, the fiber becomes as known as a leaky or hollow dielectric waveguide due to the fact that no total internal reflection exists, but rather Fresnel reflection occurs at the cladding external interface, and the portion of the light that is refracted to the surrounding material is lost from the fibre.

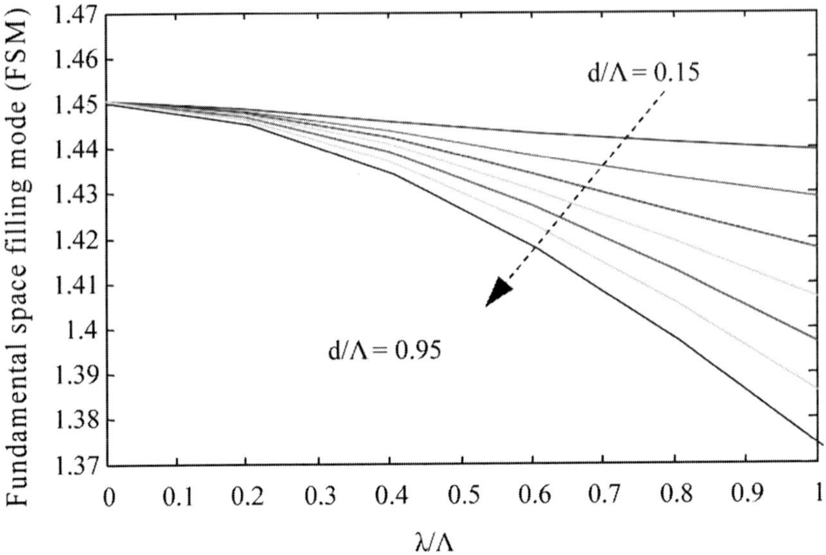

(a)

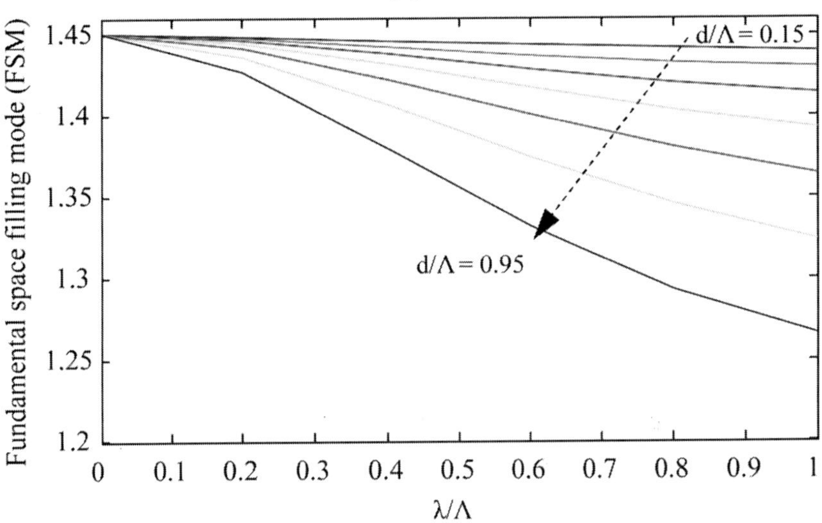

(b)

Figure 3. Effective indices for the core top (a), and fundamental space filling mode (FSM)-(b) modes for triangular photonic crystal fibers. The lines indicate different values of the hole diameter relative to the pitch. The lines correspond to different values of d/Λ, increasing by 0.1.

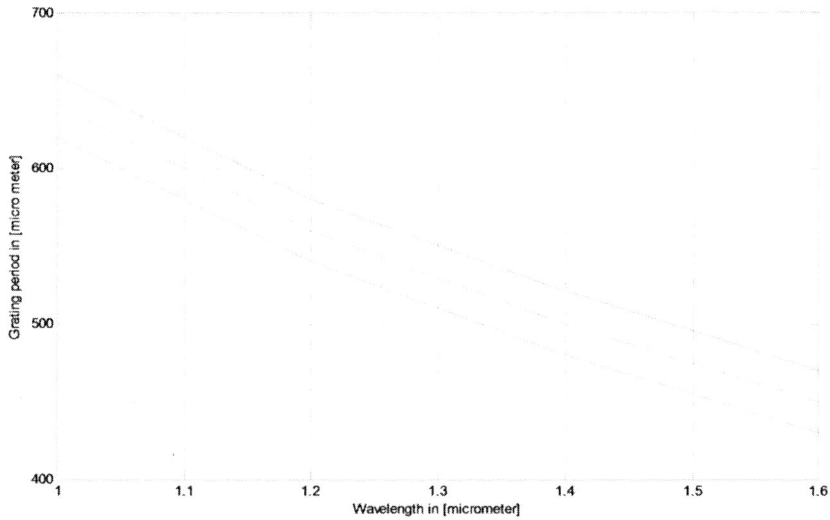

Figure 4. Grating period versus wavelength using IEIM.

3. MODAL ANALYSIS OF THE PCFS BY MMP

The CUDOS MOF software [22] is used to found both core and cladding mode by solving the transversal wave equation for the electric field:

$$\nabla\times\nabla\times \boldsymbol{E}(\boldsymbol{r})=\frac{\omega^2}{c^2}\varepsilon(\boldsymbol{r})\boldsymbol{E}(\boldsymbol{r})$$
$$\text{where } \boldsymbol{E}(\boldsymbol{r},t)=\boldsymbol{E}(\boldsymbol{r})e^{j\omega t} \tag{3}$$

The grating period of the LPGs are taken to have 500 micrometer. On the other hand to obtain a stable solution for the cladding mode we have place the radius of the air layer contain around the cladding to the minimal value that gave stable solutions [34]. In developing this model we have assumed that the dispersion due to the core, cladding and its surrounding material are negligible and therefore have been ignored. In fact all modes are guided mostly through the silica, and the influence of the material dispersion on the effective index difference is negligible [19, 35, 36]. **Figure 5** shows the field patterns of the fundamental mode (a) and the first cladding mode for different values of n_{ext}. It is clearly noted from this figure that radiation losses increase when n_{ext}become close to the cladding refractive index. In Figures 6 and 7 we report the resonant wavelength shifts and the sensitivity of the LPG fibre by using the IEIM and the MMP methods. Unless the resonance picks formed between the core mode and the first cladding mode is matched when the external index media is air, we observe that as the external refractive index increases the resonance picks calculated by IEIM shift away from the numerical value calculated by MMP. This disagreement can be explained by the contribution of the non linear effect due to the interaction between cladding mode and external refractive index. This effect is clearly observed in Figures 6 and 7 which reflects that the ambient media plays a key role in the coupling with the cladding modes.

In fact, for LPG fibres the influence of the surrounding refractive index can be described by the following analytical expression:

$$\left(\frac{d\lambda}{dn_{ext}}\right)=\left(\frac{d\lambda}{dn_{cl}^{eff}}\right)\left(\frac{dn_{cl}^{eff}}{dn_{ext}}\right)$$

From this expression it can be easily anticipated that the spectral sensitivity presents a nonlinear behavior that is more pronounced as the surrounding refractive index become close to the cladding refractive index. This nonlinear behavior is not taken in to account in the IEIM approach which decreases the accuracy of this method.

So, the IEIM can be adopted to design a single mode fibre where the procedure of calculation of the resonance coupling between the core mode and the first cladding mode in an LPG is simple. It can also offer all information related to the performance of such sensor based on this type of symmetrical LPG.

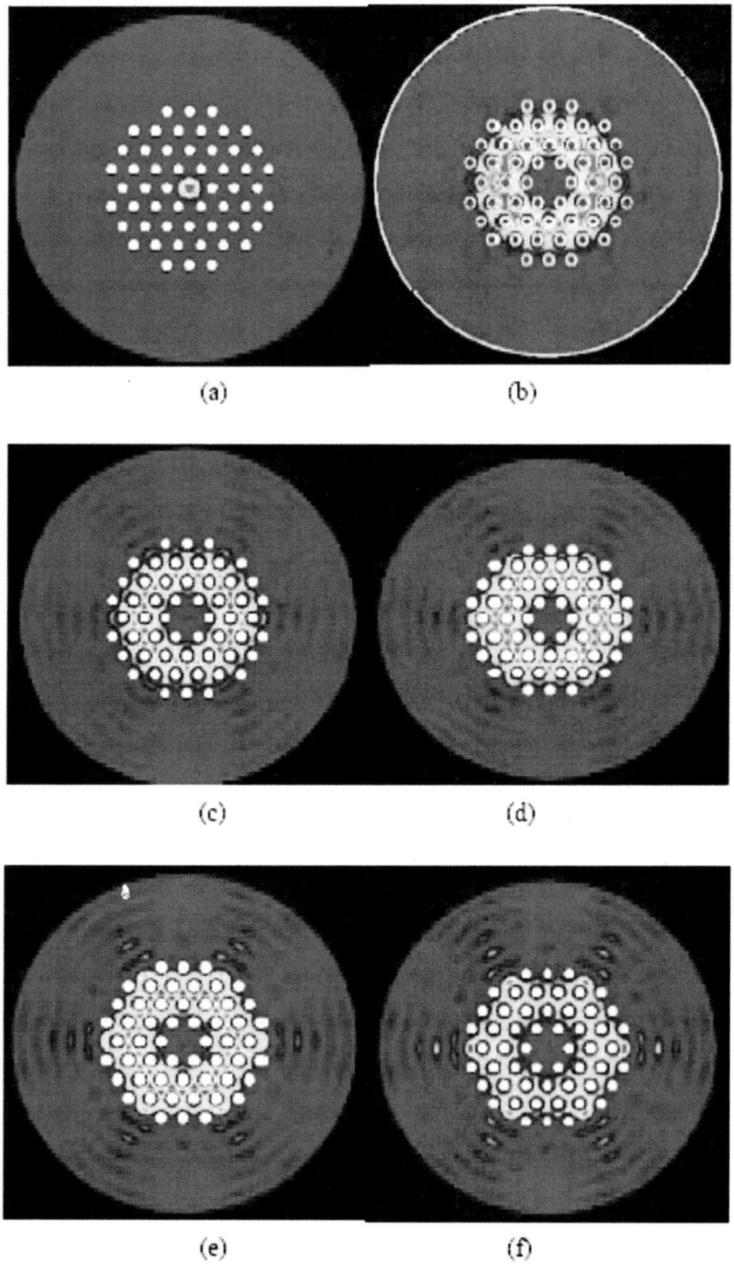

Figure 5. Profile of the first cladding mode for various external refractive index media (n_{ext}). (a); Fundamental mode, (b, c, d, e, d); cladding mode for n_{ext} = 1, 1.2, 1.3, 1.35, 1.4.

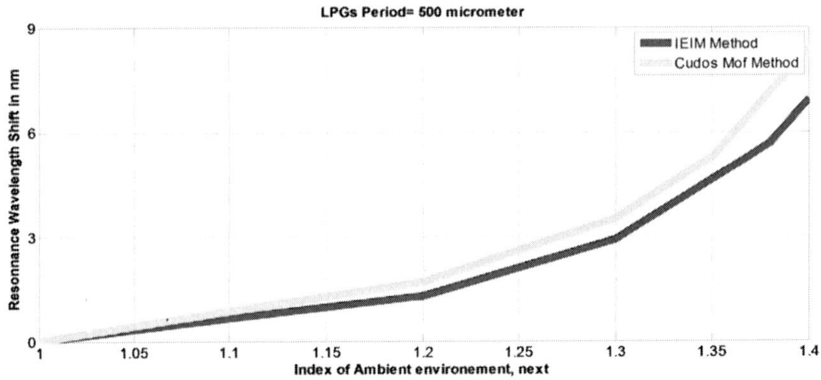

Figure 6. Calculation of the resonance shift Wavelength of the LPG-PCF.

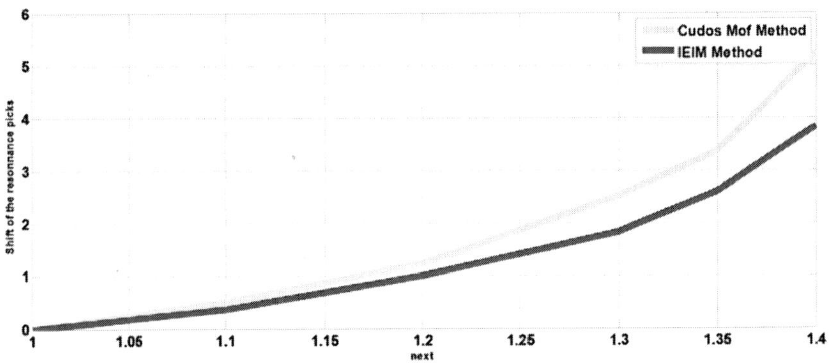

Figure 7. Sensitivity of the LPG calculated by both Cudos and IEIM method.

To further analyse the performance of the IEIM method, we have calculated the sensitivities of the proposed device for two different wavelengths, as reported in table 1. From this table, it can be anticipated that the disagreement decreases by reducing the operating wavelength of the LPG-Sensor. Next, we have investigated the sensitivity of the fibre against hole diameter in order to understand its effect on the accuracy of IEIM method. The simulating results are reported in figure 8, from which, we can conclude that results obtained by last method become less accurate when the air hole diameter decreases. This can be explained by the fact that the values of the PCF radius vary as function of both d/Λ and the relative wavelength d, Λ.

4. CONCLUSION

In conclusion, the results of this work show that IEIM is a simple and promising method for obtaining modal characteristics and sensing properties of long period grating photonic crystal fibers (LPG-PCFs) when the surrounding media is assumed to be air. However, its accuracy decreases when the fiber is immersed into a liquid liquid having a refractive index close to that of the cladding.

Table 1. Comparison of the sensitivity calculated with the two methods for the proposed LPG sensors against different wavelengths.

ESM-12-01	Refractive index sensing			
	Calcul with MM		Calcul with IEIM	
	γ_{RI}	Δλ/Δn [nm]	γ_{RI}	Δλ/Δn [nm]
LPG, 400 nm	8.24	12.67×10^3	6.73	7.8×10^3
LPG, 500 nm	20.64	29.84×10^3	15.89	18×10^3

LMA 20	Refractive index sensing			
	Calcul with MM		Calcul with IEIM	
	γ_{RI}	Δλ/Δn [nm]	γ_{RI}	Δλ/Δn [nm]
LPG, 750 nm	2.78	5.90×10^3	2.19	2600
LPG, 900 nm	11.37	20.46×10^3	9.21	14×10^3

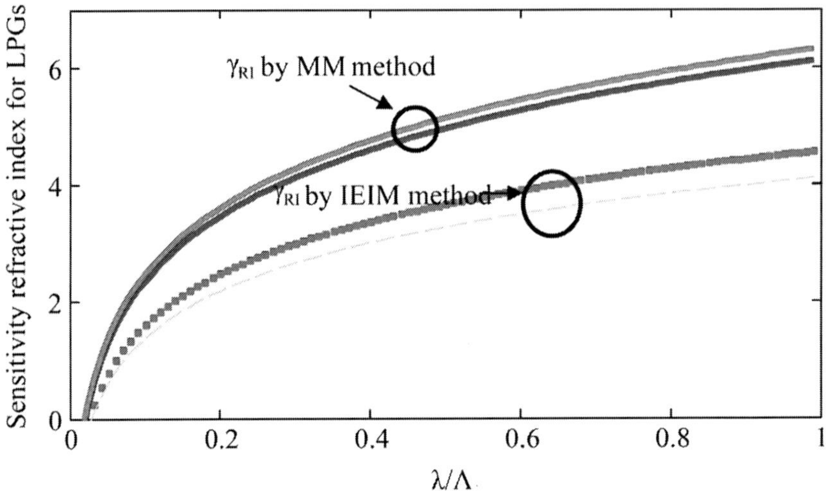

Figure 8. γ_{RI} for LPGs refractive index sensor. The lines indicate different values of the hole diameter relative to the pitch: 0.46 and 0.47 for wavelength 1 μm.

ACKNOWLEDGEMENTS

The authors would like to thank Dr. Boris Kuhlmey and Dr. Maksim Skorobogatiy, for their contribution to this work.

REFERENCES

1. H. A. Sotelo, Yu. O. Barmenkov and A. V. Kir'yanov, "The Use of Erbium Fiber Laser Relaxation Frequency for Sensing Refractive Index and Solute Concentration of Aqueous Solutions," Laser Physics Letters, Vol. 5, No. 11, 2008, pp. 825-829.

2. V. Bhatia, M. K. Burford, K. A. Murphy and A. M. Vengsarkar, "Long-Period Fiber Grating Sensors," OFC'96 Optical Fiber

Communications, San Jose, 25 Feburary-1 March 1996, pp. 265-266.

3. C. C. Ye, S. W. James and R. P. Tatam, "Simultaneous temperature and Bend Sensing with Long-Period Fiber Gratings," Optics Letters, vol. 25, No. 14, 2000, pp. 1007- 1009.

4. V. Bhatia, D. Campbell and R. O. Claus, "Simultaneous strain and Temperature Measurement with Long-Period Gratings," Optics Letters, vol. 22, No. 9, 1997, pp. 648- 650.

5. H. J. Patrick, C. C. Chang, and S. T. Vohra, "Long period Fobre Gratings for Structural Bend Sensing," Electronics Letters, vol. 34, No. 18, 1998, pp. 1773-1775.

6. W. Belhadj, F. AbdelMalek and H. Bouchriha, "Characterization and study of Photonic Crystal Fibres with bends," Materials Science & Engineering C, Vol. 26, No. 2-3, 2006, pp. 578-579.

7. X. Shu and D. Huang, "Highly Sensitive Chemical Sensor Based on the measurement of the separation of Dual Resonant Peaks in a 100-μm-period fiber grating," Optics Communications, vol. 171, No. 1-3, 1999, pp. 65-69.

8. T. Allsop, L. Zhang and I. Bennion, "Detection of Organic Aromatic Compounds in paraffin by a Long-Period Fiber Grating Optical Sensor with Optimized Sensitivity," Optics Communications, vol. 191, No. 3-6, 2001, pp. 181-190.

9. R. Falate, R. C. Kamikawachi, M. Müller, H. J. Kalinowski and J. L. Fabris, "Fiber Optic Sensors for Hydrocarbon Detection," Sensors and Actuators B: Chemical, vol. 105, No. 2, 2005, pp. 430-436.

10. H.-P. Loock, R. S. Brown, J. A. Barnes, N. R. Trefiak, K. L. Laugesen and G. Nemova, "Long Period Grating Sensor Methods and Apparatus," US Patent no. 7391942, 2004.

11. J. H. Chong, P. Shum, H. Haryono, A. Yohana, M. K. Rao, C. Lu and Y. N. Zhu, "Measurements of refractive Index Sensitivity Using Long-Period Grating Refractometer," Optics Communications, Vol. 229, No. 1-6, 2004, pp. 65-69.

12. L. Rindorf and O. Bang, "Highly Sensitive Refractometer with a Photonic-Crystal-Fiber Long-Period Grating," Optics Letters, Vol. 33, No. 6, 2008, pp. 563-565.

13. D. K. C. Wu, B. T. Kuhlmey and B. J. Eggleton, "Ultrasensitive Photonic Crystal Fiber Refractive Index Sensor," Optics Letters, Vol. 34, No. 3, 2009, pp. 322-324.

14. G. Kakarantzas, T. Birks and P. Russell, "Structural longPeriod Gratings in Photonic Crystal Fibers," Optics Letters, Vol. 27, No. 12, 2002, pp. 1013-1015.

15. Y. Zhu, P. Shum, H.-J. Chong, M. K. Rao and C. Lu, "Strong resonance and a Highly Compact Longperiod Grating in a Large-Mode-Area Photonic Crystal Fiber," Optics Express, Vol. 11 No. 16, 2003, pp. 1900-1905.

16. K. N. Park, T. Erdogan, K. S. Lee, "Cladding Mode Coupling in Long-Period Gratings Formed in Photonic Crystal Fibers," Optics Communications, Vol. 266, No. 2, 2006, pp. 541-545.

17. L. Rindorf and O. Bang, "Sensitivity of Photonic Crystal Fiber Grating Sensors: Biosensing, Refractive Index, Strain, and Temperature Sensing," Journal of the Optical Society of America B, Vol. 25, No. 3, 2008, pp. 310-324.

18. F. Brechet, J. Marcou, D. Pagnoux and P. Roy, "Complete analysis of the characteristics of propagation into photonic Crystal Fibers by the Finite Element Method," Optical Fiber Technology, Vol. 6, No. 2, 2000, pp. 181- 191.

19. H. P. Uranus and H. J. W. M. Hoekstra, "Modeling of microstructured waveguides using a finite-element-based vectorial Mode Solver with Transparent Boundary Conditions," Optics Express, Vol. 12, 2004, pp. 2795-2809.

20. H. P. Uranus and H. J. W. M. Hoekstra, "Modelling of Microstructured Waveguides Using a Finiteelement-Based Vectorial Mode Solver with Transparent Boundary Conditions," Optics Express, Vol. 12, No. 12, 2004, pp. 2795- 2809.

21. Y.-f. Li, C.-y. Wang and M.-l. Hu, "A Fully Vectorial Effective Index Method for Photonic Crystal Fibers: application to Dispersion Calculation," Optics Communications, vol. 238, No. 1-3, 2004, pp. 29-33.

22. CUDOS, "The University of Sydney ARC Centre of Excellence, School of Physics," www.physics.usyd.edu.au/cudos/mof-software

23. K. wang N. Park, T. Erdogan, K. yung and S. Lee, "Cladding Mode Coupling in long Period Gratings Formed in Photonic Crystal Fibers," Optics Communications, Vol. 266, No. 2, 2006, pp. 541-545.

24. K. N. Park and K. S. Lee, "Improved Effective-Index Method for analysis of Photonic Crystal Fibers," Optics Letters, Vol. 30, No. 9, 2005, pp. 958-960.

25. T. Erdogan and J. Lightwave, "Fiber grating spectra," Journal of Lightwave Technology, Vol. 15, No. 8, 1997, pp. 1277-1294.

26. S. W. James and R. P. Tatam, "Optical Fibre Long-Period Grating Sensors: Characteristics and application," Measurement Science and Technology, Vol. 14, No. 5, 2003, pp. R49-R61.

27. A. M. Vengsarkar, P. J. Lemaire, J. B. Judkins, V. Bhatia, T. Erdogan and J. E. Sipe, "Long Period Fiber Gratings as Band-Rejection

Filters," Journal of Lightwave Technology, Vol. 14, No. 1, 1996, pp. 58-64.

28. V. Bhatia, "Applications of Long-Period Gratings to single and Multi-Parameter Sensing," Optics Express, Vol. 4, No. 1, 1999, pp. 457-466.

29. H. J. Patrick, A. D. Kersey and F. Bucholtz, "Analysis of the response of Long Period Fiber Gratings to external index of refraction," Journal of Lightwave Technology, Vol. 16, No. 9, 1998, pp. 1606-1612.

30. Y. n. Zhu, Z. h. He and H. Du, "Detection of External Refractive Index Change with High Sensitivity Using Long-Period Gratings in Photonic Crystal Fiber," Sensors and Actuators B, Vol. 131, No. 1, 2008, pp. 265-269.

31. I. Del Villar, M. Achaerandio, I. R. Matias and F. J. Arregui, "Deposition of overlays by electrostatic Self-Assembly in Long-Period Fiber Gratings," Optics Letters, Vol. 30, No. 7, 2005, pp. 720-722.

32. Blaze Photonics, Endlessly Single Mode PCF: ESM-12- 01. http://www.blazephtonics.com/pdf/ESM%20-%2012%20-% 2001.pdf

33. J. S. Petrovic, H. Dobb, V. K. Mezentsev and K. Kalli "Sensitivity of LPGs in PCFs Fabricated by an Electric Arc to Temperature, Strain, and External Refractive Index," Journal of Lightwave Technology, Vol. 25, No. 5, 2007, pp. 1306-1312.

34. P. R. McIsaac, "Symmetry-Induced Modal Characteristics of Uniform Waveguides—I: Summary of Results," IEEE Transactions on Microwave Theory and Techniques, Vol. 23, No. 5, 1975, pp. 421-429.

35. T. W. MacDougall, S. Pilevar, C. W. Haggans and M. A. Jackson, "Generalized expression for the growth of Long Period Gratings,"

IEEE Photonics Technology Letters, vol. 10, no. 10, 1998, pp. 1449-1451.

36. X. Shu, L. Zhang and I. Bennion, "Sensitivity characteristic of long Period Fibre Gratings," Journal of Lightwave Technology, vol. 20, no. 2, 2002, pp. 255-266.

9

LOW-LOSS COUPLING OF QUANTUM CASCADE LASERS INTO HOLLOW-CORE WAVEGUIDES WITH SINGLE-MODE OUTPUT IN THE 3.7–7.6 MM SPECTRAL RANGE

Pietro Patimisco [1,2], Angelo Sampaolo [1,2], Laura Mihai [3], Marilena Giglio [1], Jason Kriesel [4], Dan Sporea [3], Gaetano Scamarcio [1], Frank K. Tittel [2] and Vincenzo Spagnolo [1,*]

[1] Dipartimento Interateneo di Fisica, Università degli studi di Bari Aldo Moro e Politecnico di Bari, Via Amendola 173, Bari I-70126, Italy

[2] Department of Electrical and Computer Engineering, Rice University, 6100 Main Street, Houston, TX 77005, USA

[3] National Institute for Laser, Plasma and Radiation Physics Laser Metrology and Standardization Laboratory 409 Atomistilor, 077125 Magurele-Bucharest, Romania

[4] Opto-Knowledge Systems, Inc. (OKSI), 19805 Hamilton Ave., Torrance, CA 90502-1341, USA

ABSTRACT

We demonstrated low-loss and single-mode laser beam delivery through hollow-core waveguides (HCWs) operating in the 3.7–7.6

μm spectral range. The employed HCWs have a circular cross section with a bore diameter of 200 μm and metallic/dielectric internal coatings deposited inside a glass capillary tube. The internal coatings have been produced to enhance the spectral response of the HCWs in the range 3.5–12 μm. We demonstrated Gaussian-like outputs throughout the 4.5–7.6 μm spectral range. A quasi single-mode output beam with only small beam distortions was achieved when the wavelength was reduced to 3.7 μm. With a 15-cm-long HCW and optimized coupling conditions, we measured coupling efficiencies of >88% and transmission losses of <1 dB in the investigated infrared spectral range.

KEYWORDS

single-mode output; hollow-core waveguide; quantum cascade laser; optical coupling

1. INTRODUCTION

Many imaging and sensing applications require a good laser beam quality with circular symmetry, as close as possible to the fundamental transverse TEM_{00} mode, and essentially diffraction-limited low beam divergence with M^2 values in the range 1.0–1.3 [1,2,3]. Commercial laser sources are often packaged with internal optics providing collimated output beams, which define and stabilize the output laser mode. These are typically designed and optimized to provide beam divergence of a few milliradians. However, the beam profile often does not result in a single-mode Gaussian-like shape, and typically exhibits an elliptical profile. Spatial defects, such as beam tails, can be easily removed by means of a pinhole. The most common beam collimating system is composed of an aspheric lens, which focuses the laser beam through a pinhole, and collimating optics, which collects the light passing through the pinhole [4,5]. If the laser beam contains higher-order modes, a modal filter for beam cleaning is essential. The ideal modal filter rejects all input field components that have no overlap with the fundamental mode of the filter, without attenuation of this mode [6]. Chalcogenide glass fibers are the most used single-mode solid-core optical waveguides in the infrared spectral range. They have a broad infrared wavelength transmission window between 1.5 and 11 μm. They are usually based on a step-index configuration and are designed to be single mode for wavelengths above a selected cut-off wavelength value. The coupling efficiency has a maximum at a specific wavelength and drops rapidly in the remaining operational spectral range. In the chalcogenide glass family, those based on As-S-Se exhibit single-mode delivery in the near infrared region (<5.5 μm) with and a minimum transmission loss of approximately 0.1 dB/m but suffering a coupling efficiency as low as 60%–70% [7]. For the infrared spectral range >5 μm, photonic crystal fibers have been used to provide single-mode delivery at 3.39 μm, 9.3 μm, and 10.6 μm [8], with losses around 2 dB/m at 10.6 μm [9].

Hollow-core waveguides (HCWs) can also act as efficient modal filters in the mid-IR range [10,11]. HCWs are optical fibers capable of single-mode mid-IR beam delivery [12,13] and are composed of a cylindrical glass capillary tube and metallic/dielectric layers deposited inside a hollow core. The spectral response of HCWs is determined by the thickness and the surface quality of the dielectric layer deposited inside the tube. The guided mode with lowest losses is a hybrid mode characterized by an optical power distribution corresponding to a Gaussian-like profile. Higher-order modes show losses that are larger than the fundamental one when traveling through the fiber. Hence, a minimum fiber length will exist that allows only the fundamental mode transmission through the fiber [12,13,14]. In this limit, the output beam exiting from the HCW will have an approximate Gaussian intensity distribution. Minimum loss condition and single-mode output beam can be achieved by optimizing the laser beam coupling into the HCW. This can be realized by using a lens that focuses the laser beam onto the HCW entrance. Experimentally, HCWs with bore diameters $d \sim 30$ times the wavelength have been shown to provide mid-IR laser beam delivery with low losses and single-mode output. [10]. In particular, HCWs with a bore size of d= 300 μm have been successfully employed in the range 7.6–11 μm for single-mode operation and optical losses of <1 dB/m have been measured [15,16,17]. Single-mode operation in the spectral range 5.1–10.5 μm was also demonstrated, employing HCWs with a bore size of 200 μm and internal layers specifically designed to reduce internal losses in the 5–12 μm spectral range [18,19].

In this work, we investigated the beam-guiding performances of HCWs fabricated by OptoKnoweldge Systems Inc. (OKSI) with a bore size of 200 μm and internal coatings realized to enhance the spectral response at 3.5 μm. To verify the possibility to achieve HCWs single-mode operation down to wavelengths shorter than 5 μm, we employed five quantum cascade lasers (QCLs) with emissions in the spectral range 3.7–7.6 μm and three HCWs with different lengths in the range 15–50 cm. We coupled the HCWs with the QCL sources by using appropriate focusing lenses. With optimized coupling conditions, single-mode and Gaussian-like outputs throughout the 4.5–7.6 μm range were observed, while quasi single-mode transmission, with only small spatial distortions, was achieved at 3.7 μm. In the last part of the manuscript, we compared the optical performance of a HCW with a bore size of 300 μm, with that of a HCW with a bore size of 200 μm, by coupling both with a 7.6 μm QCL beam. We demonstrated that, using a HCW with a 300 μm bore size, it is not possible to achieve single-mode operation at wavelength $\lambda \leq$ 7.6 μm [17].

2. EXPERIMENTAL SETUP

Three HCWs with a circular bore, having a cross-section diameter d = 200 μm and lengths of 15 cm, 30 cm, and 50 cm, produced by OKSI, were used in this work. Fabrication of the HCWs is accomplished using a wet chemistry process developed by Harrington *et al.* [12]. A silver (Ag) layer is deposited inside the glass capillary tube by flowing a silver solution through the tube using a peristaltic pump. After which, a dielectric layer is formed by flowing an iodine solution that reacts with the silver to form AgI. The longer the iodine solution flows, the thicker the resulting dielectric layer becomes. By controlling the thickness of the AgI dielectric layer, the transmission window of the fiber can be optimized for a specific wavelength range. For the HCWs employed in this work, both the reflective layer and the dielectric layer were optimized for shorter wavelengths than previously demonstrated for single-mode hollow fibers [18,19]. Modifications to the standard Ag coating technique were employed to produce a smoother surface and minimize scattering, which scales as $1/\lambda^2$. The AgI coating was designed to provide low propagation losses in the mid-infrared spectral range from 3.5 μm to 12 μm, as confirmed by the absorbance profile shown in Figure 1.

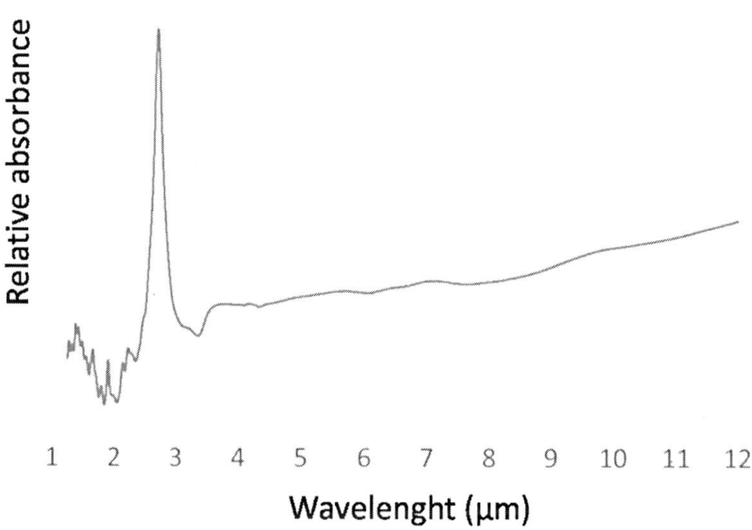

Figure 1. Hollow-core waveguide (HCW) relative absorbance measured in the range 1–12 μm using a FTIR spectrometer.

The optical coupling between the QCL and the HCW was carried out by employing the experimental setup schematically represented in Figure 2.

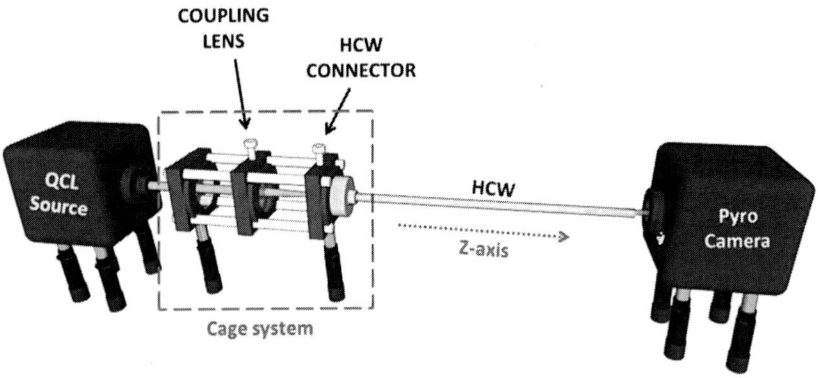

Figure 2. Schematic of the experimental setup used to optically couple quantum cascade laser (QCL) sources and the HCWs. A ZnSe lens is used to focus the collimated beam exiting from the QCL onto the waveguide entrance. An infrared pyrocamera detects the profile of the beam exiting from the HCW.

In order to investigate the beam guiding properties of HCWs in a wide mid-infrared spectral range, we employed four commercial tunable, external cavity QCLs (Pranalytica) operating at 3.7 μm (model MONOLUX 38), 4.5 μm (model MONOLUX 45), 4.9 μm (model MONOLUX 53), and 7.3 μm (model MONOLUX 74), respectively. All Monolux QCLs are equipped with custom-designed ZnSe aspherical collimating lenses, which are optimized to provide low output beam divergence. In order to realize fine adjustments of the optical coupling between the laser beam and the HCW, we developed a cage system, which includes a mount that allows a fine adjustment of the coupling lens position in the *x*–*y* planes and a fiber connector mounting to hold the HCW, as shown in Figure 2. The lens has a diameter of 1/2", and the cage system allows fine adjustments to the distance between the lens and the HCW entrance along the optical *z*-axis. The HCW connector allows changes of the HCW position entrance with respect to the focused laser spot. With such an optomechanical system, both the azimuthal and polar angles of the waveguide entrance with respect to the laser beam direction can be precisely adjusted. The far field mode profile of the laser beam exiting the HCW was recorded by means of an infrared pyro-camera (Pyrocam III, Ophir Spiricon, 124 × 124 pixels, spatial resolution of 100 μm), positioned at a distance of about 2 cm from the HCW exit. The optical power of the laser beam exiting from the HCW was measured by replacing the pyrocamera with an infrared power meter detector.

3. OPTICAL COUPLING CONDITIONS

The coupling efficiency of a fiber can be calculated from the overlap integrals between the guided modes and the laser beam [20]. The intensity distribution profile of the laser beam at the waveguide entrance can be approximated by two-dimensional (2D), Gaussian elliptical function G(*x*,*y*) given by:

$$G(x,y) = \frac{1}{2\pi\sigma_{0,x}\sigma_{0,y}} e^{-\frac{x^2}{2\sigma_{0,x}^2}} e^{-\frac{y^2}{2\sigma_{0,y}^2}} \tag{1}$$

where (x_0, y_0) are the spatial coordinates of the center, and $\sigma_{0,x}$ and $\sigma_{0,y}$ are the *x* and *y* spreads at the waveguide entrance, respectively. When the 2D laser beam enters the fiber, it transforms into a composition of the fiber modes. Hence, a coupling efficiency can be calculated for each *n*-th guided mode as:

$$\eta_n = \frac{\left[\int\limits_{-\infty}^{+\infty}\int\limits_{-\infty}^{+\infty} G(x,y)E_n(x.y)dxdy\right]^2}{\int\limits_{-\infty}^{+\infty}\int\limits_{-\infty}^{+\infty} G^2(x,y)dxdy \int\limits_{-\infty}^{+\infty}\int\limits_{-\infty}^{+\infty} E_n{}^2(x.y)dxdy} \tag{2}$$

where $E_n(x,y)$ is the n-th waveguide mode represented by a Bessel function. The lowest-order guided mode $E_1(x,y)$ has a Gaussian spatial profile with circular symmetry. As stated previously, to achieve the optimum coupling conditions, as much of the available laser beam power as possible must be coupled in the lowest-order mode. Higher-order modes suffer increased losses, and the power coupled into these modes is rapidly dissipated as the beam propagates through the waveguide. If the HWC fiber is long enough, only the $E_1(x,y)$ mode survives, and a single-mode output at the waveguide exit is obtained [20]. Numerical calculations have shown that coupling efficiencies η_1 higher than 0.89 can be obtained when both the ratios $w_{0,x}/d$ and $w_{0,y}/d$ are in the 0.5–0.75 range, where $w_{0,x} = 2\sigma_{0,x}$ and $w_{0,y} = 2\sigma_{0,y}$ are the beam diameters at the waveguide entrance. In our case, this implies that the laser beam at the waveguide entrance must have both $w_{0,x}$ and $w_{0,y}$ in the range 100–150 μm. Hence, a coupling lens with an adequate focal length must be selected to achieve this condition. Measurements of beam diameters as small as 100 μm cannot be resolved by available infrared cameras. Therefore, the beam widths $w_{0,x}$ and $w_{0,y}$ of the focused laser beam at the waveguide entrance was estimated by measuring the diameters w_x andw_y of the laser beam at the coupling lens using the following expression:

$$w_{0,i} = \frac{w_i}{\sqrt{1+\left(\frac{\pi w_i^2}{\lambda R}\right)^2}}, \text{ with } i = x, y \tag{3}$$

where R is the radius of curvature of the coupling lens. Starting from the laser beam profiles on the lens, the beam diameters w_x and w_y can be extracted by evaluating the second-order moments of the beam intensity distribution, as reported in [11].

4. BEAM PROFILES AND RELATED SPATIAL QUALITY

The QCLs beam profiles were recorded by positioning the infrared pyro-camera 7 cm away from the QCL exit and are shown in Figure 3a–d. In Table 1, we report both the measured w_x and w_y beam diameters and the calculated $w_{0,x}$ and $w_{0,y}$values obtained using Equation (3), based on the coupling lens selected for each QCL source.

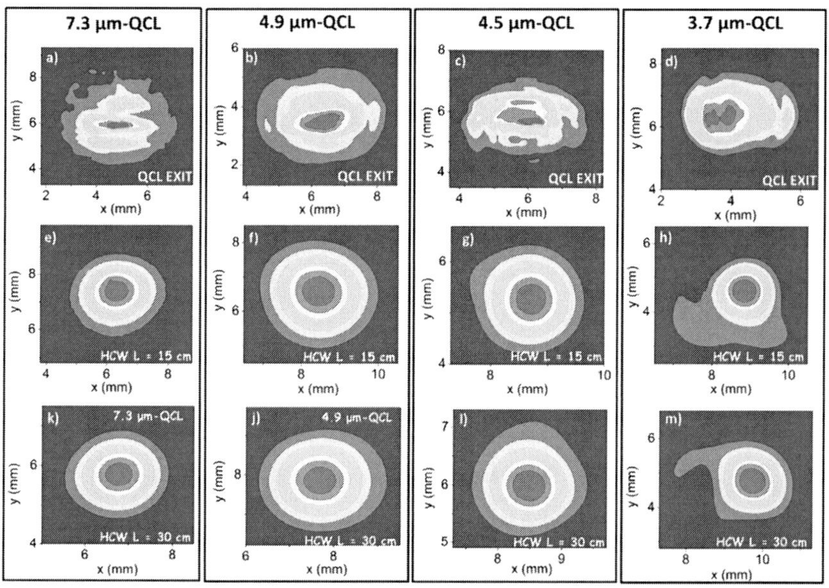

Figure 3. (**a–d**) Beam profiles of the QCL at 7.3 µm (**a**), at 4.9 µm (**b**), at 4.5 µm (**c**), and at 3.7 µm (**d**), recorded by positioning the infrared pyro-camera 7 cm away from the QCL exit. (**e–m**) The HCW output beam profiles were measured for the 15-cm- (**e–h**) and 30-cm-long (**k–m**) waveguides. The focusing lens was positioned at a distance of 7 cm from the QCL output, and the pyrocamera was located at ~2 cm from the HCW exit.

Table 1. w_x and w_y values calculated by using the second-order moments method; $w_{0,x}$and $w_{o,y}$ values calculated from Equation (3) when a coupling lens with a focal length *f*is employed.

	7.3-µm QCL	4.9-µm QCL	4.5-µm QCL	3.7-µm QCL
w_x (mm)	5.26	4.24	4.19	4.27
w_y (mm)	4.79	3.97	3.48	3.67
f (mm)	50	75	75	75
$w_{0,x}$ (µm)	132	114	96	100
$w_{0,y}$ (µm)	144	122	120	116

For our experiments, a lens with focal length f = 50 mm was selected for the QCL operating at 7.3 μm and f = 75 mm for the other three QCLs, since these f values provide $w_{0,x}$ and $w_{0,y}$ dimensions falling in the range 96–144 μm, nearly identical to the optimal one, as described above. We coupled the QCL beams into the waveguide by positioning the focusing lens 7 cm from the QCL output. The best coupling conditions were obtained by maximizing the HCW output power. Figure 3e–m show the beam profiles obtained at the exit of a 15-cm- and 30-cm-long HCWs in straight positions for the QCL emitting at 7.3 μm (Figure 3e,k), at 4.9 μm (Figure 3f,j), 4.5 μm (Figure 3g,l), and 3.7 μm (Figure 3h,m). To collect the beam profile, we positioned the pyro-camera ~2 cm far from the waveguide exit.

The measured profiles demonstrated that, despite the low quality of the input laser beam, HCWs with a core size as small as 200 μm allow single-mode output with a Gaussian-like beam profile down to a wavelength of 4.5 μm. Indeed, the QCLs mode are perfectly matched to the $E_1(x,y)$ guided mode, as shown by the circular-symmetric beam output [21]. Similar single-mode results were obtained for the 50-cm-long HCW. For the shorter wavelength QCL emitting at 3.7 μm, the acquired profiles show a Gaussian-like pattern with small tails. However, only <3% of the total power is contained in the beam tails, which can be removed using a pinhole. A similar profile was also observed for the 50-cm-long fiber.

The HCWs output beam spatial quality can be quantified by calculating the parameter M^2 which compares the laser beam angular divergence in the two transverse directions with a diffraction-limited Gaussian beam, which has M^2 = 1. For the beam exiting from the QCLs, the M^2 value can be defined as:

$$M^2_{i,QCL} = \frac{\theta_{i,QCL}}{2}\frac{\pi w_i}{\lambda}, \text{ with } i = x, y \tag{4}$$

where $\lambda/(\pi\ w_i)$ is the beam divergence half-angle of a diffraction-limited Gaussian beam, and $\theta_{x(QCL)}$ and $\theta_{y(QCL)}$ are the divergence half-angles of the QCL beams. The M^2 value of the beam exiting from the HCWs can be defined as the ratio between the half-angle beam divergence of the beam and the theoretical half-angle beam divergence:

$$M^2_{i,HCW} = \frac{\theta_{i,HCW}}{2} \frac{\pi d}{2.4048 \cdot \lambda}, \text{ with } i = x, y \tag{5}$$

where $2.4048 \cdot \lambda/(\pi \cdot d)$ is the theoretical half-angle divergence, calculated by assuming that only the $E_1(x,y)$ guided mode exits from the waveguide. For both, the QCLs and the waveguide outputs, half-angle beam divergences of the real beam can be calculated by extracting the beam widths w_x and w_y from the beam profiles acquired at different distances using the experimental approach reported in [22]. Results obtained for the output of all four QCLs together with results for the same beams after exiting the 15-cm-long HCW are listed in Table 2.

The calculated M^2 values confirm that the laser beam exiting from the L = 15 cm HCW shows a substantial improvement in spatial quality with respect to the input laser beam. Similar M^2 values were obtained for the 30-cm- and 50-cm-long HCWs.

5. TOTAL LOSSES AND COUPLING EFFICIENCY

The propagation losses Lp of an HCW can be calculated by using the following expression:

$$L_P(dB) = -10\log_{10}\left(\sum_n \eta_n e^{-2\alpha_n L}\right) \tag{6}$$

Table 2. Divergence angles and M^2 values for four QCLs and the 15-cm-long HCW, in two directions (*x* and *y*) orthogonal to the QCL beam propagating *z*-direction.

	7.3-μm QCL	4.9-μm QCL	4.5-μm QCL	3.7-μm QCL
$\theta_{x,QCL}$ (mrad)	2.1	1.9	1.5	1.3
$\theta_{y,QCL}$ (mrad)	2.0	1.7	2.2	1.1
$M^2_{x,\,QCL}$	4.75	5.16	4.39	4.71
$M^2_{y,\,QCL}$	4.12	4.32	4.85	3.42
$\theta_{x,HCW}$ (mrad)	30.72	22.48	21.97	20.48
$\theta_{y,HCW}$ (mrad)	31.53	22.22	22.30	19.92
$M^2_{x,\,HCW}$	1.12	1.18	1.26	1.44
$M^2_{y,\,HCW}$	1.15	1.17	1.28	1.40

where α_n are the attenuation coefficients related to the n-th guided modes. These coefficients depend on the laser wavelength and the optical properties of the dielectric and metallic layers deposited inside the HCW. For their estimation, we used the relation provided by Miyagi and Kawakami [23]:

$$\alpha_n = \left(\frac{u_n}{2\pi}\right)^2 \frac{8\lambda^2}{d^3}\left(\frac{n_m}{n_m{}^2 - k_m{}^2}\right)\left\{\frac{1}{2}\left[1 + \frac{n_d^2}{\left(n_d^2 - 1\right)^{1/2}}\right]^2\right\} \tag{7}$$

where n_m and k_m are the real and imaginary parts of the complex index of the silver layer, n_d is the complex index of the silver iodine layer, and u_n is the n-th root of the zero-order Bessel function. The values of n_m, k_m, and n_d were calculated by using relations reported in [24]. Experimental values of the HCW total losses can be calculated from the ratio between the power at the waveguide entrance I_0 and that measured at the fiber exit I_S as:

$$Losses(dB) = 10\log_{10}\left(\frac{I_0}{I_S}\right) \tag{8}$$

Figure 4a–d show the total losses measured for the 15-cm-, 30-cm- and 50-cm-long HCWs, using QCLs emitting at 7.3 μm (Figure 4a), 4.9 μm (Figure 4b), 4.5 μm (Figure 4c), and 3.7 μm (Figure 4d), for their respective best coupling conditions.Figure 4 also includes the theoretical losses, calculated by using Equations (6) and (7), as a function of the HCWs length and assuming that the waveguide modes $E_n(x,y)$, with n ranging from 1 to 5, propagate inside the waveguide.

Measured optical losses are higher than the calculated ones. These discrepancies can be attributed to the poor spatial laser beam quality, which impacts: (i) the estimation of theoretical coupling efficiencies (calculated by assuming a perfect two-dimensional Gaussian elliptical profile for the laser beam) and the consequent propagation losses; and (ii) the beam waist sizes at the waveguide entrance. From the intercept of the linear fit in Figure 4, we extracted the coupling efficiency values η_{exp}, reported in Table 3, together with the theoretical values calculated by using Equations (1) and (2). A very good data agreement (<7% discrepancy) was obtained.

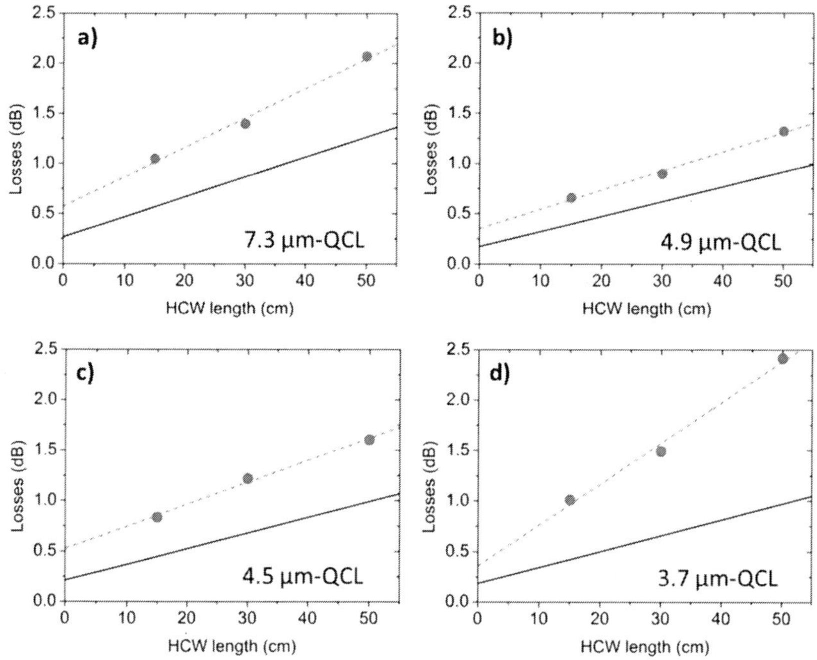

Figure 4. (**a–d**) Total losses (•) measured when coupling the 7.3-μm- (**a**), 4.9-μm- (**b**), 4.5-μm- (**c**) and 3.7-μm-QCL (**d**) beams into the HCWs having lengths of 15, 30, and 50 cm, plotted as a function of the HCW length. Dashed lines are linear fits to the data. Solid lines are the theoretical losses calculated by using Equations (6) and (7).

In Figure 5, we report the total losses, plotted as a function of the lasers wavelength for all three HCW lengths.

Table 3. Theoretical coupling efficiencies into the lowest-order mode η_1 and experimental η_{ext} values.

	7.3-μm QCL	4.9-μm QCL	4.5-μm QCL	3.7-μm QCL
η_1	0.94	0.96	0.95	0.96
η_{exp}	0.88	0.92	0.89	0.92

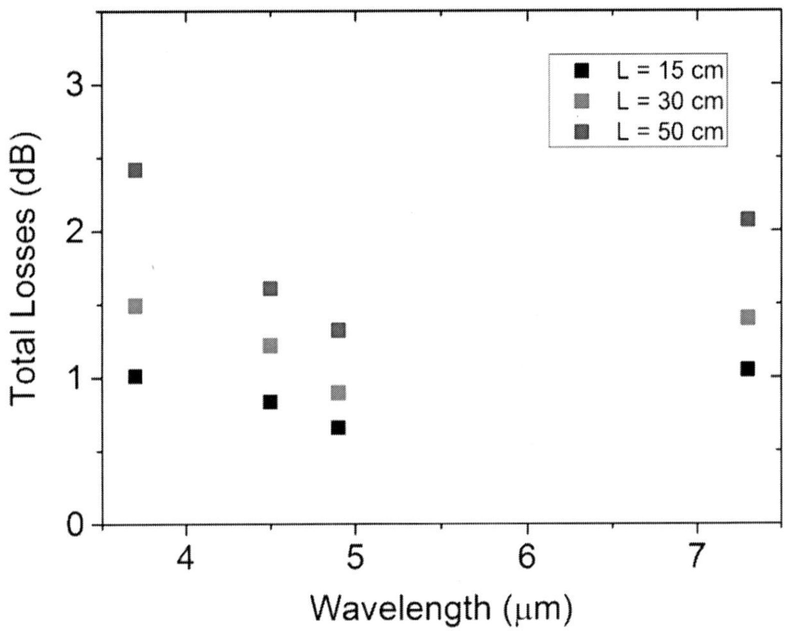

Figure 5. **(a–d)** Total losses as a function of the QCLs emitting wavelengths for HCWs with L = 15 cm (■), 30 cm (■), and 50 cm (■).

This graph clearly shows that, for all investigated HCW lengths, the measured loss changes are less than 1 dB when the laser wavelengths span from 3.7 to 7.3 μm, confirming that the internal coatings enhance the HCW spectral response to 3.7 μm. Moreover, the losses remain ≤1 dB in the investigated wavelength range for the 15-cm-long HCW.

6. INFLUENCE OF BORE DIAMETER AND HCW LENGTH ON THE OUTPUT BEAM QUALITY

The $1/d^3$ dependence (Equation (7)) predicts a strong increase in HCW losses as the fiber bore diameter is reduced. A large bore diameter cannot ensure a single-mode profile at the waveguide output. From Equation (6), we can see that for large bore sizes α_n is small and higher-order modes have lower losses with respect to those obtained for small bore sizes, for which α_n is larger. For example, calculations of attenuation coefficients using Equations (7) and (8) at λ = 7.6 μm and L = 1 m show that, for d = 200 μm, losses related to the second-order mode (n = 2) are 11.3 dB, but decreases to 3.3 dB when d is 300 μm. Thus, there is significantly less mode mixing at the HCW output in waveguides with reduced bore size. Mode mixing can be reduced by increasing the fiber length at the expense of larger optical losses. The spatial quality of the input laser beam must also be taken into account, since it influences the input power percentage coupled into the guided modes (see Equation (2)). The larger the amount of input power coupled into the high-order modes, the larger their contributions to total propagation losses will be. As a result, a HCW with a small bore size can exhibit a single-mode output profile, but with optical losses significantly higher than those theoretically predicted. To investigate the influence of the HCW bore size on optical losses and output beam profile, we coupled a 200-μm- and 300-μm-core-sized HCWs, both having a length of 15 cm, with a Daylight Solution external cavity QCL (DLS-QCL) emitting at 7.6 μm. The beam profile at the DLS-QCL exit is shown in Figure 6a, characterized by w_x = 2.56 mm and w_y = 1.79 mm. The DLS-QCL was coupled with a 200-μm-core-sized HCW by using a coupling lens with a focal length of 25 mm (resulting in $w_{0,x}/d$ = 0.48 and $w_{0,y}/d$ = 0.67). For the 300-μm-core-sized HCW, a focal length of 50 mm ($w_{0,x}/d$ = 0.63 and $w_{0,y}/d$ = 0.88) was used. The output profiles of the 200-μm- and 300-μm-core-sized HCWs (both with L = 15 cm) are shown in Figure 6b,c, respectively.

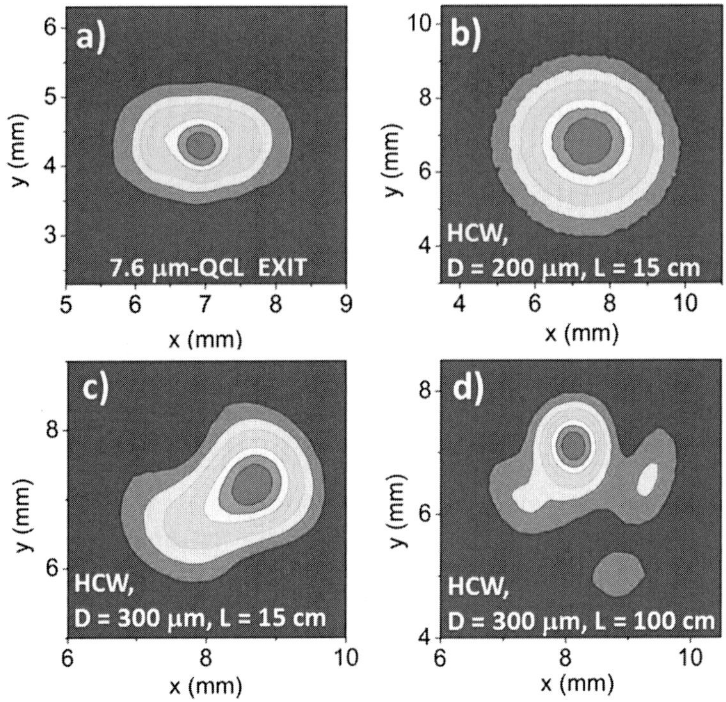

Figure 6. (**a**) Beam profile at the exit of DLS-QCL emitting at 7.6 μm. (**b–d**) Beam profiles with a 200-μm-core-sized and 15-cm-long HCW (**b**); 300-μm-core-sized and 15-cm-long HCW (**c**); and 300-μm-core-sized and 100-cm-long HCW (**d**) exits. The focusing lens was positioned at a distance of 5 cm from the QCL output and the pyrocamera was located at ~2 cm from the HCWs exit.

The intensity distribution at the output of the 200-μm-core-sized and 15-cm-long HCW, shown in Figure 6b, is single mode. At the HCW exit, we measured optical losses of 0.74 dB, slightly higher than those theoretically predicted (0.65 dB) by using Equations (6) and (7). This confirms that the quality of the laser beam influences the total losses, but not the output beam profile. Conversely, the intensity distribution is multimodal for the 300-μm-core-sized and 15-cm-long HCW, as shown in Figure 6c. We measured losses as high as 1.14 dB, 0.56 dB higher than the theoretical value. This discrepancy is largely due to the propagation of higher-order modes, which have greater losses than the lowest-order E_1 mode. To verify whether the contribution to higher-order modes at the 300-μm-core-sized HCW exit can be suppressed by increasing the fiber length, we coupled the DLS-QCL with a 100-cm-long, 300-μm-core-sized HCW. The coupling conditions were the same as used for 300-μm-core-sized HCW. The acquired output beam profile is reported in Figure 6d. Clearly, the profile is still multimodal, with optical losses of 3.45 dB (1.62 dB are the theoretical losses), demonstrating that the only way to achieve single-mode beam delivery in the spectral range of 4.5–7.6 μm is to use HCWs with a bore size as small as 200 μm. For this condition, a length as short as 15 cm is sufficient to provide a single-mode output, even if the spatial quality of the input laser beam is not good. For longer fibers, single-mode output is ensured but optical losses increase. For the spectral range 8–11 μm, both 200-μm and 300-μm diameters can guarantee single-mode beam delivery [16,17,18], but it is preferable to employ a bore size of 300 μm, since it provides lower optical losses.

7. CONCLUSIONS

Prior to this work, single-mode laser beam delivery in the mid-infrared range was demonstrated at $\lambda \leq 5.5$ μm with solid core fibers [7,25] and at $\lambda \geq 5.1$ μm by using HCWs [18]. In this work, despite the limited beam quality of the input laser beams, we demonstrated single-mode propagation down to 4.5 μm and nearly single-mode at 3.7 μm with only slight spatial distortions through HCWs with bore diameters of 200 μm. Thus, it was shown that there are single-mode fiber options covering the 3.7 μm–12 μm mid-IR spectral range. For the 15-cm-long HCW, we measured straight optical losses of ≤1 dB and a coupling efficiency of ≥88% in the investigated 3.7–7.3 μm spectral range. In all cases, the measured losses are higher than those theoretically predicted, and the discrepancies can be attributed to the limited quality of the input QCL beams. We also demonstrated that, using HCWs with a bore diameter of 300 μm, it is not possible to achieve single-mode operation for $\lambda \leq 7.6$ μm (40 times lower than the bore size). Otherwise, 200-μm-core HCWs are capable of operating in single-mode down to wavelength 50 times lower than their bore sizes, which, to date, represents the largest d/λ ratio ever observed for HCWs.

ACKNOWLEDGMENTS

Part of this work was supported by the COST Action MP1204 "TERA-MIR radiation: materials, generation, detection and applications." The authors from the Dipartimento Interateneo di Fisica di Bari acknowledge financial support from Italian research projects PON02 00675 and PON02 00576 and PON03 "SISTEMA". F. K. Tittel acknowledges support by the Robert Welch Foundation (grant C-0586) and a NSF ERC MIRTHE award. The Romanian authors acknowledge the financial support of the Romanian Space Agency (ROSA), under the ECSA project, grant 67/2013. The QCLs used in this research were purchased in the frame of the project "Center for Advanced Lasers Technologies (CETAL)" (Romania), contract 8PM/2010, financed by UEFISCDI, Romania.

AUTHOR CONTRIBUTIONS

P.P, A.S., M.G., J.K, G.S., F.K.T. and V.S. conceived and designed the experiments; P.P., A.S., L.M., and M.G. performed the experiments and analyzed the data; D.S., and L.M. provided the QCL sources; J.K. provided the hollow core waveguide samples. All wrote the paper.

CONFLICTS OF INTEREST

The authors declare no conflict of interest.

REFERENCES

1. Patimisco, P.; Scamarcio, G.; Tittel, F.K.; Spagnolo, V. Quartz-Enhanced Photoacoustic Spectroscopy: A Review.*Sensors* **2014**, *14*, 6165–6206.

2. Gagliardi, G.; Loock, H.P. *Cavity-Enhanced Spectroscopy and Sensing*; Springer: London, UK, 2014.

3. Willer, U.; Saraji, M.; Khorsandi, A.; Geiser, P.; Schade, W. Near- and mid-infrared laser monitoring of industrial processes, environment and security applications. *Opt. Laser. Eng.* **2006**, *44*, 699–710.

4. Hariharan, P. *Optical Holography*; Cambridge University Press: Cambridge, UK, 1996.

5. Kermene, V.; Saviot, A.; Vampouille, M.; Colombeau, B.; Froehly, C.; Dohnalik, T. Flattening of the spatial laser beam profile with low losses and minimal beam divergence. *Opt. Lett.* **1992**, *17*, 859–861.

6. Ksendzov, A.; Lay, O.; Martin, S.; Sanghera, J.S.; Busse, L.E.; Kim, W.H.; Pureza, P.C.; Nguyen, V.Q.; Aggarwal, I.D. Characterization of mid-infrared single mode fibers as modal filters. *Appl. Opt.* **2007**, *46*, 7957–7962.

7. Houizot, P.; Boussard-Plédel, C.; Faber, A.J.; Cheng, L.K.; Bureau, B.; Van Nijnatten, P.A.; Gielesen, W.L.M.; Pereira do Carmo, J.; Lucas, J. Infrared single mode chalcogenide glass fiber for Space. *Opt. Express* **2007**, *15*, 12529–12538.

8. Caillaud, C.; Renversez, G.; Brilland, L.; Mechin, D.; Calvez, L.; Adam, J.-L.; Troles, J. Photonic bandgap propagation in all-solid chalcogenide microstructured optical fibers. *Materials* **2014**, *7*, 6120–6129.

9. Butvina, L.N.; Sereda, O.V.; Dianov, E.M.; Lichkova, N.V.; Zagorodnev, V.N. Single-mode microstructured optical fiber for the middle infrared. *Opt. Lett.* **2007**, *32*, 334–336.

10. Kriesel, J.M.; Hagglund, G.M.; Gat, N.; Spagnolo, V.; Patimisco, P. Spatial mode filtering of mid-infrared (mid-IR) laser beams with hollow core fiber optics. *Proc. SPIE* **2013**, *8993*.

11. Patimisco, P.; Sampaolo, A.; Giglio, M.; Kriesel, J.M.; Tittel, F.K.; Spagnolo, V. Hollow core waveguide as mid-infrared laser modal beam filter. *J. Appl. Phys.* **2015**, *118*, 113102–113106.

12. Harrington, J.A. *Infrared Fibers and Their Applications*; SPIE: Bellingham, WA, USA, 2004.

13. Kriesel, J.M.; Gat, N.; Bernacki, B.E.; Erikson, R.L.; Cannon, B.D.; Myers, T.L.; Bledt, C.M.; Harrington, J.A. Hollow Core Fiber Optics for Mid-Wave and Long-Wave Infrared Spectroscopy. *Proc. SPIE* **2011**, *8018*.

14. Bledt, C.M.; Harrington, J.A.; Kriesel, J.M. Loss and modal properties of Ag/AgI hollow glass waveguides. *Appl. Opt.* **2012**, *51*, 3114–3119.

15. Spagnolo, V.; Patimisco, P.; Borri, S.; Scamarcio, G.; Bernacki, B.E.; Kriesel, J. Part-per-trillion level SF_6 detection using a quartz enhanced photoacoustic spectroscopy-based sensor with single-mode fiber-coupled quantum cascade laser excitation. *Opt. Lett.* **2012**, *37*, 4461–4463.

16. Spagnolo, V.; Patimisco, P.; Borri, S.; Scamarcio, G.; Bernacki, B.E.; Kriesel, J. Mid-infrared fiber-coupled QCL-QEPAS sensor. *Appl. Phys. B* **2013**, *112*, 25–33.

17. Siciliani de Cumis, M.; Viciani, S.; Borri, S.; Patimisco, P.; Sampaolo, A.; Scamarcio, G.; De Natale, P.; D'Amato, F.; Spagnolo, V. A widely-tunable mid-infrared fiber-coupled quartz-enhanced photoacoustic sensor for environmental monitoring. *Opt. Express* **2014**, *22*, 28222–28231.

18. Sampaolo, A.; Patimisco, P.; Kriesel, J.M.; Tittel, F.K.; Scamarcio, G.; Spagnolo, V. Single mode mid-IR hollow fibers operating in the range 5.1–10.5 μm. *Opt. Express* **2015**, *23*, 195–204.

19. Patimisco, P.; Sampaolo, A.; Kriesel, J.M.; Tittel, F.K.; Scamarcio, G.; Spagnolo, V. Mid-IR quantum cascade laser mode coupling in hollow-core, fiber-optic waveguides with single-mode beam delivery. *Proc. SPIE* **2015**, *9370*.

20. Nubling, R.; Harrington, J.A. Launch conditions and mode coupling in hollow glass waveguides. *Opt. Eng.* **1998**, *37*, 2454–2458.

21. Patimisco, P.; Spagnolo, V.; Vitiello, M.S.; Tredicucci, A.; Scamarcio, G.; Bledt, C.M.; Harrington, J.A. Coupling external cavity mid-IR quantum cascade lasers with low loss hollow metallic/dielectric waveguides. *Appl. Phys. B***2012**, *108*, 255–260.

22. Patimisco, P.; Spagnolo, V.; Vitiello, M.S.; Tredicucci, A.; Scamarcio, G.; Bledt, C.M.; Harrington, J.A. Low-Loss Hollow Waveguide Fibers for Mid-Infrared Quantum Cascade Laser Sensing Applications. *Sensors* **2013**, *13*, 1329–1340.

23. Miyagi, M.; Kawakami, S. Design theory of dielectric-coated circular metallic waveguides for infrared transmission. *J. Lightw. Technol.* **1984**, *2*, 116–126.

24. George, R.; Harrington, J.A. Infrared transmissive, hollow plastic waveguides with inner Ag–AgI coatings. *Appl. Opt.***2005**, *44*, 6449–6455.

25. Mid-Infrared Optical Fiber, Fiber Patch Cables, and Bundles. Available online: https://www.thorlabs.com/newgrouppage9.cfm?objectgroup_id=7062#SingleMode (accessed on 12 April 2016).

10

OPTICAL EFFECTS ACCOMPANYING THE DYNAMICAL BRAGG DIFFRACTION IN LINEAR 1D PHOTONIC CRYSTALS BASED ON POROUS SILICON

Anton Maydykovskiy, Vladimir Novikov, Sergey Svyakhovskiy and Tatiana Murzina *

Physics Department, Moscow State University, 119991 Moscow, Russia

ABSTRACT

We survey our recent results on the observation and studies of the effects accompanying the dynamical Bragg diffraction in one-dimensional photonic crystals (PhC). Contrary to the kinematic Bragg diffraction, the dynamical one considers a continuous interaction between the waves travelling within a spatially-periodic structure and is the most pronounced in the so called Laue geometry, leading to a number of exciting phenomena. In the described experiments, we study the PhC based on porous silicon or porous quartz, made by the electrochemical etching of crystalline silicon with the consequent thermal annealing. Importantly, these PhC are approximately hundreds of microns thick and contain a few hundreds of periods, so that the experiments in the Laue diffraction scheme are available. We discuss the effect of the temporal splitting of femtosecond laser pulses and show that the effect is quite sensitive to the polarization and the phase of a femtosecond laser pulse. We also show the experimental realization of the Pendular effect in porous quartz PhC and demonstrate the experimental conditions for the total spatial switching of the output radiation between the transmitted and diffracted directions. All described effects are of high interest for the control over the light propagation based on PhC structures.

KEYWORDS

dynamical Bragg diffraction; photonic crystals; pulse splitting; Pendular effect

1. INTRODUCTION

Effects of light propagation through photonic crystals (PhC) have been intensively studied in last decades. They reveal a number of exciting phenomena that originate generally from the spatial periodicity of the PhC structure. First of all one should mention here the dispersion modulation in definite frequency-angular spectral intervals, that are close to the edges of the so called photonic band gap [1,2]. This leads to such well known effects as slow light, *i.e.*, a substantial decrease of the light group velocity [3], spatial and spectral light localization [4], the decrease of the optical excitation threshold of PhC based laser structures [5,6]. An important property of photonic crystals is also a manifold amplification of the efficiency of the nonlinear-optical effects such as harmonics' generation [7,8] and nonlinear magneto-optical effects [9]. When the nonlinear interaction of the pulsed laser radiation with the PhC structure is sufficient, the so-called gap solitons can exist within the photonic crystals [10,11]. These peculiarities of light-PC interaction are prospective for the development of new methods for the operation over propagation of the light flow.

A specific area in the context of the light-PhC interaction is the dynamical Bragg diffraction of the electromagnetic waves in a spatially periodic medium, which originates from a strong coupling between the coherent diffracted waves existing in a PhC, often named Borrmann and anti-Borrmann modes and which are localized within the layers with low and high refractive indices, respectively. It may be realized for the Laue transmission scheme, when the electromagnetic radiation propagates in the direction parallel to the PC layers under the fulfillment of the Bragg diffraction conditions. It is known from the X-ray research area that in this experimental geometry the so-called pendulum effect can exist, which is a periodic oscillation of the energy between the transmitted and diffracted waves in a PhC [12,13]. This effect is of special interest for an all-optical switching of laser pulses propagating inside a PhC. Borrmann effect is also a well-known phenomenon observed for the dynamical Bragg diffraction conditions for the X-rays [14]. Recently, a perspective optical phenomenon of diffraction-induced laser pulses splitting (DIPS) under the dynamical Bragg diffraction in the Laue scheme of diffraction in a linear one-dimensional (1D) PC was described theoretically [15,16]. It was shown that short (femtosecond) laser pulses after passing through a PhC can be split into two, the separation time being determined by the PhC properties.

In this paper we discuss our recent results on the experimental observation and studies of the pendulum and DIPS effects along with their pronounced polarization selectivity in the Laue geometry in 1D photonic crystals.

2. DYNAMICAL BRAGG DIFFRACTION IN LINEAR 1D PHC: THE MAIN APPROACH

The optical effects discussed below for the Laue geometry of the Bragg diffraction came mostly from the X-ray spectral range. The corresponding theory was also developed, which could be to some extend applied to the optical wavelength range. Still the following quite important differences should be clearified. The most important one is that the value of the variation of the refractive index modulation for the case of photonic crystals is quite large (can be over 10%), that is approximately four orders of magnitude higher as compared to the X-rays spectral range for the electronic crystals. One of the important circumstances of this property, as was shown in [15], is a high spatial inhomogeneity of the distribution of optical field within a PC, so that a single incident short laser pulse splits into two that propagate in a photonic crystal with different group velocities, that was named the diffraction induced pulse splitting (DIPS).

Importantly, that the fast (Borrmann) pulse is formed by the two coupled diffracted waves that are localized in PC layers with smaller refractive index, while the slow (anti-Borrmann) pulse is formed by the interacting waves that are concentrated predominantly within the layers with high refractive index. This is shown schematically in Figure 1a. In the case of a large enough pulse propagating length inside a PC, the splitting time between the Borrmann and of the anti-Borrmann pulses can be large enough and be detected in the experiment.

Theoretical description of the dynamical Bragg diffraction was developed recently (see [15,16] and references therein). It is based on solving the boundary problem of the interacting fields for the Laue geometry and for the spatially confined laser pulses. Briefly, the incident laser pulse at the PC surface (*i.e.*, at $z = 0$) is taken in the form of a two-dimensional Fourier expansion, that is as a series of monochromatic plane waves that correspond to the frequency interval $[\omega_0 \pm \Omega]$, where ω_0 is the central frequency, $\Omega << \omega_0$, and with the amplitudes that are the functions of the coordinates z, x and of the time t; the coordinate system is shown in Figure 1a as well. In the frame of the two-wave approximation and under the fulfillment of the Bragg diffraction condition, $\sin \vartheta_B = \lambda/2d$, d being the PC period, the p-polarized electric field within the bulk of the photonic crystal is determined by the expression [16]:

$$\mathbf{E}(x,z,t)=\{\mathbf{E}_0(x,z,t)+\mathbf{E}_h(x,z,t)\exp(-iGx)\}\times\exp(ik_{0x}x-i\omega_0 t) \quad (1)$$

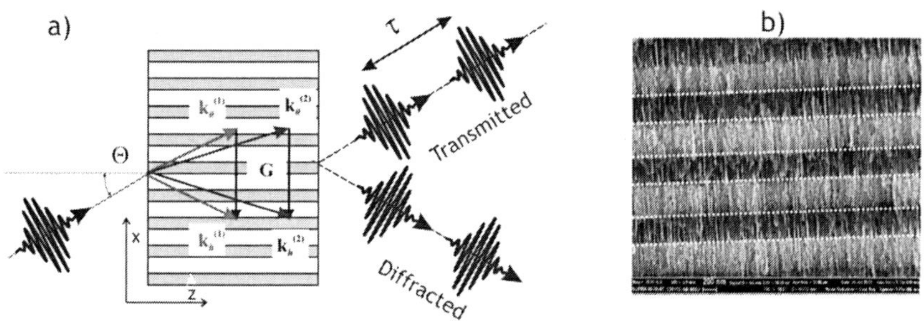

Figure 1. (**a**) Scheme of the dynamical Bragg diffraction and of the temporal pulse splitting in the direct and diffraction directions; (**b**) SEM image of the cross-section of the porous silicon based PhC.

where $G = 2\pi/d$ is the reciprocal lattice vector, $\mathbf{E}_0\ (x, z, t)$ and $\mathbf{E}_h\ (x, z, t)$ are the amplitudes of the transmitted and diffracted waves, respectively; the wave vector $k = \omega/c = k_{0x} + K$, $k_{0x} = k_0 \sin\theta$ is the lateral projection of the wave vector, θ is the angle of incidence. In other words, a coherent superposition of the incident, E_0, and of the diffracted, E_h, pulse fields determine the field in every point within the PhC structure. This two-wave approximation is valid in particular due to an achievable large modulation of the permittivity (of the refraction index) of the layers that compose the photonic crystal. The permittivity of the PhC in this approximation is:

$$\varepsilon(x)=\varepsilon_0+\varepsilon_h\exp(-i\,hx)+\varepsilon_{-h}\exp(i\,hx) \tag{2}$$

where ε_0, ε_h, ε_{-h} are the spatial Fourier components of the permittivity, the exact equations that determine their dependence on the PhC material and geometrical parameters of a particular structure are described in [16]. Importantly, the amplitudes of the confined fields $\mathbf{E}_0$ (x, z, t) and $\mathbf{E}_h$ (x, z, t) are given by integrating over the frequency-angular distribution of the propagating pulses, thus giving the dependence of the output field on both the dielectric parameters of the photonic crystal, as well as on the duration, phase, amplitude and propagation direction of the incident pulsed radiation.

3. COMPOSITION OF MULTILAYER POROUS SILICON BASED 1D PHOTONIC CRYSTALS

In this work we performed the experiments for the 1D PhC based on porous quartz. The technique of the composition of such structures is well known and is described in detail elsewhere (see, for example, [17] and references therein). It is based on the electrochemical etching of crystalline (001) *p*-type silicon of the resistivity of about 0.005 Ω cm in HF-based electrolyte (HF:H_2O:ethanol in the ratio 2:4:3 for the HF concentration of 21% (w/w)), which results in the formation of pores oriented along the [100] direction, *i.e.*, perpendicularly to the surface of the Si(001) plate. During this procedure, the silicon plate serves as the bottom electrode, while the platinum wire is the second electrode in a two-electrode electrochemical cell. It was checked that the pores' diameter and thus the porosity of the porous silicon layer are proportional to the electrochemical current density, while the etching time determines the depth of the pores. Thus, the periodic modulation of the current density results in the formation of spatially periodic porous structure, the layers of the constant porosity being oriented parallel to the Si(001) surface. The average pores' diameter in our conditions were varied in the interval 10–60 nm [18], which is much smaller that the optical wavelength, so that such porous layers may be treated as a homogeneous media. As a result, the PC made of several hundreds of layers with low porosity (high refractive index, n_1) and of high porosity (smaller refractive index, n_2) was made, its average thickness being about 300 microns.

When fabricating the PhC structures for the experiments in the Laue geometry of the Bragg diffraction, the following requirements were considered. First, the PhC should contain a large number of layers, so that the cross-section of the structure would be large enough to carry out the optical experiments, where the laser spot diameter is of dozens of microns. Second, the optical thickness of the layers was made to satisfy the relation $n_1d_1 = n_2d_2 = \lambda/4$, where λ is the central wavelength of the photonic band gap and determines the Bragg condition and is close to 800 nm, which is the fundamental wavelength of our laser systems. As a result, a high periodic structure is formed; the cross-section is shown in Figure 1b.

Most of the optical experiments were performed when using the output radiation of a Ti-sapphire laser at 800 nm, where the absorption of (porous) silicon is large enough. In order to make the PhC structure transparent at this wavelength, we performed a thermal annealing of the structure at T = 900 °C for a couple of hours. It was proven that annealing does not break the structural periodicity of the PhC, while leads to a substantial decrease of the refractive index of the porous layers and consequently to a spectral shift of the photonic band gap and Bragg diffraction condition. Thus the corresponding calibration measurements and the necessary corrections of the thicknesses of the porous silicon layers of the composed PhC were made for a required spectral range.

4. EXPERIMENTAL RESULTS

4.1. Temporal Splitting of the Femtosecond Laser Pulses

The experiments on the temporal splitting of short laser pulses were performed for a 1D porous quartz based PhC 300 μm thick (consisting of 375 periods of 390 nm thick), the refractive indexes of the alternating layers being $n_1 = 1.45$ and $n_2 = 1.35$. The radiation of a Ti-sapphire laser at the wavelength of 800 nm and the pulse duration of 30 fs was used, which was focused on a cut-off of a PhC structure into a spot of approximately 30 μm in diameter. The diffraction Bragg condition was realized at the angle of incidence of 31°.

Prior to the experimental studies, FDTD calculations were performed in order to model the process, the parameters of the sample, of the angle of incidence of 30 fs laser pulses being the same as in the experiment. The corresponding results of the instantaneous spatial distribution of the amplitude squared of the propagating pulse are shown in Figure 2 for different time intervals after the entrance of the laser pulses into the PhC. It can be seen that for the times above approximately 400 fs the input pulse is clearly separated into two ones.

Most evidently it can be seen on the cross-sections of the 3D distribution (see Figure 2b), where we present the intensity distribution calculated for the coordinates x that correspond to the centers of layers with low and high refractive index. It stems from Figure 2b that the first ("fast") pulse from a pair is concentrated mostly in layers with low n and corresponds to the Borrmann mode. On the contrary, the second ("slow") pulse is localized predominantly in more optically dense layers and thus is associated with the anti-Borrmann mode.

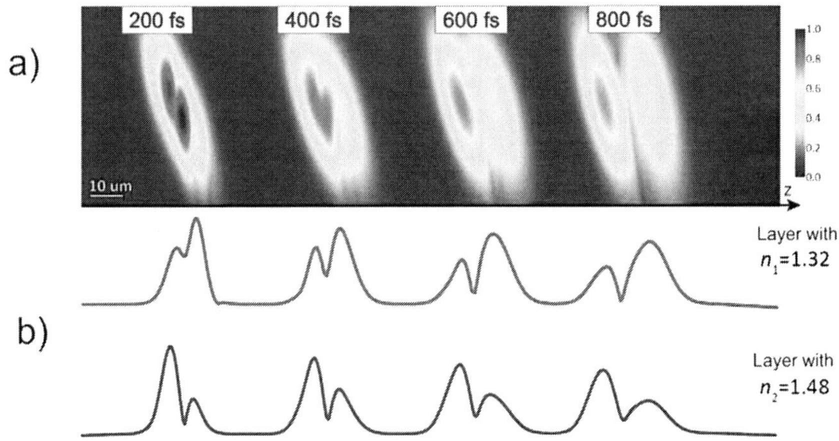

Figure 2. Finite difference time domain (FDTD) calculation of the propagation of a 30 fs laser pulse through a PhC in the Laue scheme of the dynamical Bragg diffraction; the PhC layers are parallel to the *z*-axis (shown in panel a) and perpendicular to the plane of the figure. (**a**) Spatial distribution of the pulse intensity after the laser pulse travels for 200 fs, 400 fs, 600 fs and 800 fs inside a PhC; (**b**) cross-sections of the panel a made for the layers of low porosity (upper red curve) and of high porosity (bottom blue curve).

Experiments were performed using the cross-correlation scheme based on a Ti-sapphire laser, the pulse width in the signal and reference arms being of 30 fs. The sample was placed in the signal arm in the Laue diffraction geometry, the laser radiation was focused onto its facet into a spot of 20 μm in diameter. The cross correlation function was studied when measuring the intensity of the second harmonic generated in a BBO crystal under the noncollinear sinchronism, as the pulses passed though the PhC and the reference arms were focused on it and as the time delay τ between the pulses from the signal and the reference arms was varied: $I_{CC}(\tau) \infty \int \infty - \infty\ I_S(t)\ I_R(t + \tau)dt$, where $I_S(t)$ and $I_R(t + \tau)$ are the pulse intensities in the signal and the reference channels.

Figure 3a shows the cross-correlation functions measured for the *s*- and *p*-polarized fundamental radiation. In both cases the diffraction-induced pulse splitting (DIPS) effect is observed, while the separation time is different for the two polarizations and is τ(p)12 ≈ 746 fs for the *p*-polarized fundamental radiation and τ(p)12 ≈ 520 fs for the s-polarization.

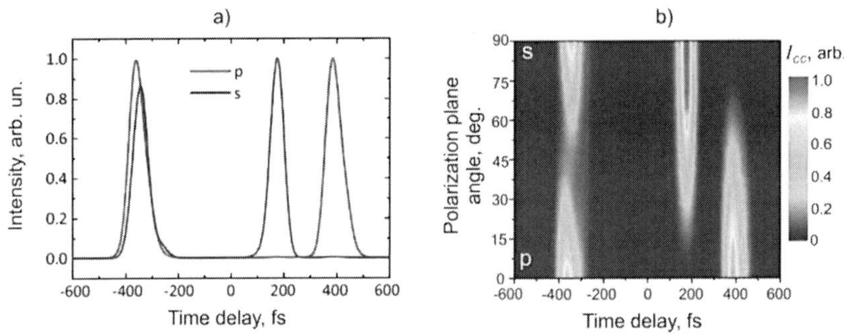

Figure 3. (**a**) Cross-correlation functions measured for the *p*- and *s*-polarized fundamental radiation and (**b**) as a function of the angle between the polarization plane of the fundamental radiation and the plane of incidence.

This same effect is evident from Figure 3b, which shows the dependence of the cross-correlation function on the polarization of the incident pulsed radiation and a continuous transformation of the output pulse splitting for the cases of the *p*- and *s*-polarizations. In particular, for the *mixed* polarization (polarization plane is tilted to 45° with respect to the plane of incidence) the output pulse is split temporally into three.

The mechanism underlying the observed polarization sensitivity of the DIPS effect, namely that the time splitting is approximately 1.5 times larger for the case of *p*-polarization, is two-fold, as was discussed in [19]. First, it is induced by the lattice anisotropy of a 1D PhC. It appears from the full expression for the wave equation if we consider, that ∇×∇×E(r,t)=−ΔE(r,t)−∇(ε−1 (∇ε)E(r,t)), *i.e.*, as the gradient of the dielectric permittivity is taken into consideration. The following expressions for the time splitting in the DIPS effect were obtained [19]:

$$\tau_{12}^{(s)} = \frac{z\,\varepsilon_h}{c\,\varepsilon_0^{1/2}}\left(1 - \frac{\sin^2 \vartheta_B}{2\varepsilon_0}\right); \quad \tau_{12}^{(p)} = \frac{z\,\varepsilon_h}{c\,\varepsilon_0^{1/2}}\left(1 + 3\frac{\sin^2 \vartheta_B}{2\varepsilon_0}\right) \tag{3}$$

where z is the PhC length and ε_0, ε_h are the zero and the first spatial Fourier components of the permittivity, and c is the speed of light.

Second, we have to take into consideration the *material* anisotropy of porous quartz. Previous studies [20,21] as well as our experiments [22] have shown that it is quite significant in our case, so that for the refractive indices for the ordinary and extraordinary beams are $n_{1,0}$ = 1.35 ± 0.01; $n_{1,e}$ = 1.32 ± 0.01 and $n_{2,0}$ = 1.46 ± 0.01; $n_{2,e}$ = 1.45 ± 0.01. The performed calculations show a good correlation with the experimental dependencies if both types of anisotropy are considered.

4.2. Dynamical Bragg Diffraction of Chirped Femtosecond Laser Pulses: The Effect of Selective Pulse Compression

Recently it was predicted that the dynamical Bragg diffraction should lead to distinct effects for the propagation of chirped pulses in a PhC, namely, compression or decompression of the pulses after passing through a PhC may be observed depending in the chirp sign [16]. Below we show the experimental evidence of this effect, first reported in [23].

The PhC used for these experiments and the cross-correlation scheme were similar to described in the previous section. The phase modulation of the laser pulse in the signal arm was introduced by a four-prism compressor, which allowed to modulate the pulse width in the interval 35–140 fs when keeping constant the pulse spectral width; the linear phase modulation (chirp) can be characterized by the parameter $\beta=\phi''(\Delta\omega)^2\approx\pm\sqrt{(\tau/\tau_0)^2-1}$. where τ_0, τ are the pulse duration prior to the compression and after it, correspondingly, ϕ'' is the second phase derivative, that is the group delay dispersion, $\Delta\omega$ is the pulse spectral width. The compressed signal pulse was focused on the PhC facet into a spot of 20 μm in diameter. The pulse duration in the reference channel was kept to be 30 fs.

Figure 4a shows the dependencies of the widths of the Borrmann and of the anti-Borrmann pulses that are observed after passing through the photonic crystal; pulse duration were estimated from the deconvolution of the measured cross-correlation functions for various values of the chirp parameter β. For the comparison, the estimated pulse duration measured without a PhC in the signal arm is shown in the same graph. It can be seen that in general the pulse widths of the first (Borrmann) and the second (anti-Borrmann) outgoing pulses for every β value are different. Namely, for $\beta < 0$ the Borrmann pulse is compressed up to the duration of 80 fs, while the laser pulse measured without the sample in the sample channel is about 96 fs. At the same time, the duration of the anti-Borrmann pulse exceeds 120 fs under the same experimental conditions.

In other words, we observe the effect of the selective pulse compression (decompression).

For positive chirp values the situation is reversed. At zero β the duration of both the Borrmann and of the anti-Borrmann pulses are equal to each other within the experimental accuracy; it exceeds the pulse width of the reference laser pulse due to the material dispersion of the photonic crystal.

The results of the FDTD calculations of the spatial widths of the Borrmann and anti-Borrmann pulses for the experimentally realised conditions are shown in Figure 4b. They reveal a good correlation with the experimental data and also testify the observation of the effect of selective compression of chirped laser pulses in the Laue scheme of the Bragg diffraction.

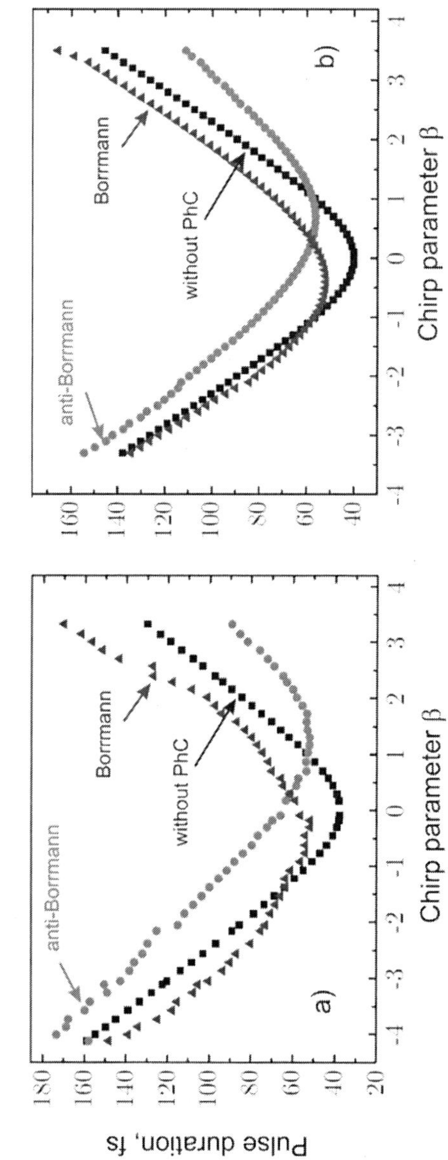

Figure 4. (**a**) Experimental and (**b**) calculated dependencies of the values of the duration of the Borrmann, anti-Borrmann and input chirped pulses on the chirp parameter β.

4.3. Pendulum Effect in the Laue Diffraction Scheme in 1D Porous Silicon PhC

The pendulum (Pendellösung) effect, which is well-known from X-ray optics, consists in a periodic exchange of energy between the transmitted and diffracted waves, which are shown schematically in Figure 1a. To some extend it is analogous to the behavior of a mechanical system consisting of two coupled pendulums; in optics it is described by the dynamical Bragg diffraction theory. Recently the pendulum effect in microwaves was studied for 2D photonic crystal made of an array of aluminum cylinders [13,24], while only a few periods of the electromagnetic field beating within a PhC were observed. A possibility of making polarizing beam splitter based of the Pendellösung effect in a photonic crystal is considered theoretically and numerically in [25], where the possibility to achieve the negative refraction in a PhC is exploited. The concept of this effect is close to the so called superprism effect in PhC [26,27]. The observation of the pendulum effect in the optical range reported in [25] for a holographic grating was still related to a relatively low contrast of the refractive index in the structure so that a small number of periods of the oscillations were achieved.

Below we describe the results on the studies of the Pendellösung effect in a 1D linear photonic crystal (PC) at the Laue diffraction scheme.

The experiments were made for a 1D linear PhC based on porous quartz and similar to described in Section 4.1, the length of the PhC sample being about 0.3 mm. An OPO laser system with a pulse duration of 10 ns and with the wavelength (idler) tuning range from 720 nm up to 1200 nm was used as the light source. In order to decrease the spectral width of the incident beam, a spatial filter was introduced in front of the sample, so that the line width of about 1 nm was used. The studies of the pendulum effect were performed close to the Bragg diffraction condition, which is at 30° angle of incidence and close to the wavelength of 800 nm. After the sample, the intensity of the transmitted and diffracted beams was detected using the photodiodes. The polarization of the incident radiation was controlled by the double Fresnel rhombus.

Figure 5a shows the spectral dependence of the intensity of the diffracted beam when *s*-polarized fundamental radiation was used. It reveals intensity oscillations with the average period of $\Delta\tau = 3.56 \pm 0.01$ nm, which is much smaller than the previously reported values. Similar spectra were measured simultaneously for the two outgoing channels (see Figure 5b), which shows that the oscillations in the transmitted and diffracted beams are in antiphase and their periods are equal to each other. This proves that the observed phenomenon is the Pendellösung effect.

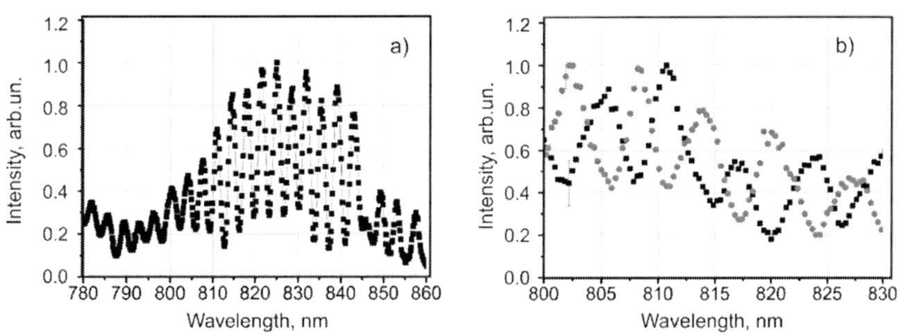

Figure 5. (**a**) Intensity spectrum of the diffracted beam passed through the porous quartz PhC of 0.2 mm long in the Laue geometry; (**b**) spectra of the transmitted (red symbols) and of diffracted (black symbols) beams under measured under the similar conditions.

It is worth noting that the intensity modulation depth of these dependencies is rather high and reaches 60%, that is quite enough for a reliable detection of the spatial switching of the laser beam between the transmitted and diffracted directions when varying the wavelength of the incident radiation. Similar dependencies were attained for the *p*-polarized fundamental beam, while the oscillation period is larger and is about 6 nm for the chosen experimental geometry [28]. This is again due to the material dispersion of the porous quartz PhC.

The period of the Pendulum effect in a PhC is determined by the expression

$$2\Lambda^{(s,p)} = 2\pi / \Delta q_z^{(s,p)} = 2\pi / \left(q_{z1}^{(s,p)} - q_{z2}^{(s,p)}\right) \qquad (4)$$

where $\Lambda^{(s,p)}$ is the extinction length and $q_{z1}^{(s,p)}$, $q_{z2}^{(s,p)}$ are the z-projections of the wave vectors of the Borrmann and of the anti-Borrmann *p*- or *s*-polarized waves within the photonic crystal. Figure 5 demonstrates the pendulum effect induced by the variation of the wavelength of light incident on the photonic crystal, while qualitatively similar dependencies can be realized for say when changing continuously the length of the PhC structure [28].

While the physics underlying the observed Pendular effect in linear 1D PhC is known, still the results shown in Figure 5are new and rather promising. In fact, a wide flexibility of the electrochemical technique that is used for the composition of porous quartz photonic crystals allows for the fabrication of an optical spatial splitter for the desired spectral range. Wide flexibility of the fabrication procedure is supported by the usage of the electrochemical etching technique in combination with the thermal annealing, which allows to vary easily both the optical and "physical" thickness of the PhC layers. Moreover, the period of the intensity redistribution between the transmitted and diffracted beams, which is proportional to the extinction parameter of the PhC, can be controlled by choosing the thickness of the PhC structure.

5. CONCLUSIONS

Summing up, we have studied a number of linear optical effects accompanying the dynamical Bragg diffraction of light in the Laue geometry in porous quartz based photonic crystals. Namely, the experimental observation of the temporal splitting of the femtosecond laser pulses and its polarization sensitivity, selective compression (decompression) of the temporally split pulses with the linear phase modulation, and the pendulum effect with a small spectral period of a few nanoseconds are discussed. The observed phenomena are very promising for the control over the propagation of the laser radiation, both for the short (femtosecond) pulses and for a continuous radiation as well.

ACKNOWLEDGMENTS

This work was in part supported by RFBR, grant No. 13-02-00300.

AUTHOR CONTRIBUTIONS

Anton Maydykovskiy, Vladimir Novikov and Sergey Svyakhovskiy have made all the experiments; Vladimir Novikov and Sergey Svyakhovskiy performed the calculations; all of them took part in the paper writing under the leadership of Tatiana Murzina.

CONFLICTS OF INTEREST

The authors declare no conflict of interest.

REFERENCES

1. Sakoda, K. *Optical Properties of Photonic Crystals*, 2nd ed.; Springer: Berlin, Germany, 2004.

2. Joannopoulos, J.D.; Johnson, S.G.; Winn, J.N.; Meade, R.D. *Photonic Crystals: Molding the Flow of Light*, 2nd ed.; Princeton University Press: Princeton, NJ, USA, 2008.

3. Gersen, H.; Karle, T.J.; Engelen, R.J.; Bogaerts, W.; Korterik, J.P.; van Hulst, N.F.; Krauss, T.F.; Kuipers, L. Real-space observation of ultraslow light in photonic crystal waveguides. *Phys. Rev. Lett.* **2005**, *94*.

4. Yeh, P.; Yariv, A.; Hong, C. Electromagnetic propagation in periodic stratified media. I. General theory. *J. Opt. Soc. Am.* **1977**, *67*, 423–438.

5. Yablonovitch, E. Inhibited spontaneous emission in solid-state physics and electronics. *Phys. Rev. Lett.* **1987**, *58*, 2059–2062.

6. Zhu, L.; Chak, P.; Poon, J.K.S.; DeRose, G.A.; Yariv, A.; Scherer, A. Electrically-pumped, broad-area, single-mode photonic crystal lasers. *Opt. Express* **2007**, *15*, 5966–5975.

7. Scalora, M.; Bloemer, M.J.; Manka, A.S.; Dowling, J.P.; Bowden, C.M.; Viswanathan, R.; Haus, J.W. Pulsed second harmonic generation in nonlinear one-dimensional periodic structures. *Phys. Rev. A* **1997**, *56*, 3166–3174.

8. Soboleva, I.V.; Murchikova, E.M.; Fedyanin, A.A.; Aktsipetrov, O.A. Second- and third-harmonic generation in birefringent photonic crystals and microcavities based on anisotropic porous silicon. *Appl. Phys. Lett.* **2007**, *87*, 241110:1–241110:3.

9. Inoue, M.; Fujikawa, R.; Baryshev, A.; Khanikaev, A.; Lim, P.B.; Uchida, H.; Aktsipetrov, O.; Fedyanin, A.; Murzina, T.; Granovsky, A. Magnetophotonic crystals. *J. Phys. D Appl. Phys.* **2006**, *39*.

10. Kivshar, Y.S.; Agrawal, G.P. *Optical Solitons: From Fibers to Photonic Crystals*; Academic: Waltham, MA, USA, 2003.

11. Mok, J.T.; de Sterke, C.M.; Eggleton, B.J. Delay-tunable gap-soliton-based slow-light system. *Opt. Express* **2006**, *14*, 11987–11996.

12. Calvo, M.L.; Cheben, P.; Martínez-Matos, O.; Monte, F.; Rodrigo, J.A. Experimental Detection of the Optical Pendellösung Effect. *Phys. Rev. Lett.* **2006**, *97*.

13. Savo, S.; di Gennaro, E.; Miletto, C.; Andreone, A.; Dardano, P.; Moretti, L.; Mocella, V. Pendellosung effect in photonic crystals. *Opt. Express* **2008**, *16*, 9097–9105.

14. Borrmann, G. Über Extinktionsdiagramme der Röntgenstrahlen von Quarz. *Phys. Z.* **1941**, *42*, 157–162. (In German).

15. Bushuev, V.A.; Mantsyzov, B.I.; Skorynin, A.A. Diffraction-induced laser pulse splitting in a linear photonic crystal.*Phys. Rev. A* **2009**, *79*.

16. Skorynin, A.A.; Bushuev, V.A.; Mantsyzov, B.I. Dynamical Bragg diffraction of optical pulses in photonic crystals in the Laue geometry: Diffraction-induced splitting, selective compression, and focusing of pulses. *J. Exp. Theor. Phys.***2012**, *115*, 56–67.

17. Pavesi, L. Porous silicon dielectric multilayers and microcavities. *Riv. Nuovo Cimento* **1997**, *20*, 1–76.

18. Svyakhovskiy, S.E.; Maydykovsky, A.I.; Murzina, T.V. Mesoporous silicon photonic structures of thousands of periods. *J. Appl. Phys.* **2012**, *112*.

19. Svyakhovskiy, S.E.; Kompanets, V.O.; Maydykovskiy, A.I.; Murzina, T.V.; Chekalin, S.V.; Skorynin, A.A.; Bushuev, V.A.; Mantsyzov, B.I. Observation of the temporal Bragg-diffraction-

induced laser-pulse splitting in a linear photonic crystal. *Phys. Rev. A* **2012**, *86*.

20. Timoshenko, V.Y.; Osminkina, L.A.; Efimova, A.I.; Golovan, L.A.; Kashkarov, P.K.; Kovalev, D.; Kunzner, N.; Gross, E.; Diener, J.; Koch, F. Anisotropy of optical absorption in birefringent porous silicon. *Phys. Rev. B* **2003**, *67*.

21. Golovan, L.A.; Melnikov, V.A.; Konorov, S.O.; Fedotov, A.B.; Timoshenko, V.Y.; Zheltikov, A.M.; Kashkarov, P.K.; Ivanov, D.A.; Petrov, G.I.; Yakovlev, V.V. Linear and nonlinear optical anisotropy of amorphous oxidized silicon films induced by a network of pores. *Phys. Rev. B* **2006**, *73*.

22. Svyakhovskiy, S.E.; Skorynin, A.A.; Bushuev, V.A.; Chekalin, S.V.; Kompanets, V.O.; Maydykovskiy, A.I.; Murzina, T.V.; Novikov, V.B.; Mantsyzov, B.I. Polarization effects in diffraction-induced laser pulse splitting in one-dimensional photonic crystals. *JOSA B* **2013**, *30*.

23. Svyakhovskiy, S.E.; Maydykovskiy, A.I.; Novikov, V.B.; Kompanets, V.O.; Chekalin, S.V.; Skorynin, A.A.; Bushuev, V.A.; Mantsyzov, B.I.; Murzina, T.V. Selective Compression of Femtosecond Laser Pulses in a Linear Photonic Crystal. In Proceedings of the Frontiers in Optics FTu3A.21, Orlando, FL, USA, 6–10 October 2013.

24. Mocella, V. Negative refraction in Photonic Crystals: Thickness dependence and Pendellosung phenomenon. *Opt. Express* **2005**, *13*, 1361–1367.

25. Mocella, V.; Dardano, P.; Moretti, L.; Rendina, I. A polarizing beam splitter using negative refraction of photonic crystals. *Opt. Express* **2005**, *13*, 7699–7707.

26. Merzlikin, A.M.; Vinogradov, A.P. Superprism effect in 1D photonic crystal. *Opt. Commun.* **2006**, *259*, 700–703.

27. Panoiu, N.C.; Bahl, M.; Osgood, R.M. Optically tunable superprism effect in nonlinear photonic crystal. *Opt. Lett.* **2003**,*28*, 2503–2505.

28. Novikov, V.B.; Svyakhovskiy, S.E.; Skorynin, A.A.; Bushuev, V.A.; Mantsyzov, B.I.; Maydykovskiy, A.I.; Murzina, T.V. Control over the Light Propagation by Means of the Pendellösung Effect in 1D Porous Quartz Photonic Crystal. In Proceedings of the ICONO/LAT 2013, Moscow, Russia, 18–22 June 2013.

THEORETICAL AND NUMERICAL STUDY OF STRESS AND THERMO-OPTIC IN PHOTONIC CRYSTAL FIBER LASER IN DIFFERENT PUMP SCHEMES

M. Abouricha[1*], A. Boulezhar[1], M. El Mouden[1,2], M. Kriraa[1], N. Rouchdi[1]

[1]Laboratory of Theoretical and Applied Physics, Faculty of Sciences-Ain Chok, Hassan II Casablanca University, Casablanca, Morocco

[2]Engineering Sciences Laboratory for Energy (LabSIPE), National School of Applied Sciences, El Jadida, Morocco

ABSTRACT

In this paper, we present a study of thermal, average power scaling, change in index of refraction and stress in photonic crystal fiber lasers with different pump schemes: forward pump scheme, backward pump scheme, forward pump scheme with reflection of 98%, backward pump scheme with reflection of 98% and bi-directional pump scheme. We show that management of thermal effects in fiber lasers will determine the efficiency and success of scaling-up efforts. In addition, we show that the most suitable scheme is the bi-directional.

KEYWORDS

Pump Schemes, PCF Lasers Birefringence in Fiber Lasers, Diode-Pumped Fiber Lasers, Fiber Lasers, Scaling of Fiber Lasers, Yb-Doped Fiber Lasers, Thermal Effects in Fiber Lasers

1. INTRODUCTION

The superiorities of the conventional solid-state and gas lasers make Yb^{3+} doped fibers lasers reported [1] -[4] . Recently a new type of the pump schemes with the high brightness semiconductor diode pump laser was studied [5] . And the new type of photonic crystal fibers (PCFs) is attracting increasing interests because of its unique properties such as endlessly single-mode guiding, freedom of dispersion characteristics, and large mode area [6] [7] . We also concentrate on PCFs in which a core doped with Yb^{3+} surrounded by a lower index cladding, which is, surrounded by an air-clad region, in turn, surrounded by a second lower index cladding index.

By analytical and numerical calculation and using the finite-difference method (FDM), we have determined the expressions of temperature distribution in different regions of the photonic crystal fiber laser (PCF) along the axial and radial directions from the integration of the steady-state heat equation for an isotropic medium. Then we used the expressions previously derived for the temperatures in Regions I, II, III, and IV [8] in the results of the expressions for the stress components ($\sigma_r(r)$, $\sigma_\varphi(r)$, and $\sigma_z(r)$) and the change in the index of refraction in Regions I, II, III, and IV. The results are compared in different pump schemes for giving the design guidelines to ensure maximum heat dissipation and saving pump powers. The calculated stress values are very small in different pump schemes, and will consequently have a negligible effect upon the index of refraction.

The resulting gradients are still small as the corresponding stress values. As a consequence, we expect that changes in the index of refraction due to the stress-optic effect will be negligible in different pump schemes, and thermally induced birefringence will be absent in fiber lasers (in all cases of pumping), the stress component $\sigma_z(r)$ increases with the increase of the length the fiber in the cases (forward pump schemes) and decreases along the fiber laser in the cases (backward pump schemes). These values are between −0.4 ´ 10^{-6} kg/m^2 and −3.4 ´ 10^{-6} kg/m^2 in the four primer cases. Moreover, in the case bi-directional pump scheme, its value is between −0.8 ´ 10^{-6}kg/m^2 and −1.8 ´ 10^{-6} kg/m^2. And the values of the change in index of refraction increases in the cases of forward pump schemes and decreases in the cases of the backward pump schemes, along the fiber laser. For the bi-directional, its value is even smaller.

2. AVERAGE TEMPERATURE IN PCF

As shown in Figure 1, a typical high power Yb^{3+}-doped double-clad PCF laser consists of an Yb^{3+}-doped dou- ble PCF with reflectors on both of the ends.

Figure 2 shows the heat flow mechanisms in fiber laser and the radial coordinate r and the tangential angle φ. The quantities a, b_1, b_2 and b are the core, inner cladding, air-clad and outer cladding radii, respectively.

The temperature distribution in a fiber reported in [8] is necessary for determining the radially varying index of refraction due to $\mathrm{d}n/\mathrm{d}T$, the calculated stresses, and the change in index of refraction through the stress-optic effect.

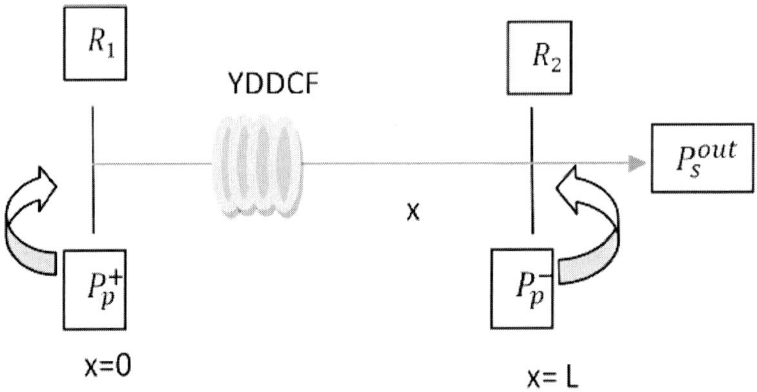

Figure 1. Schematic illustration of end pumped fiber lasers.

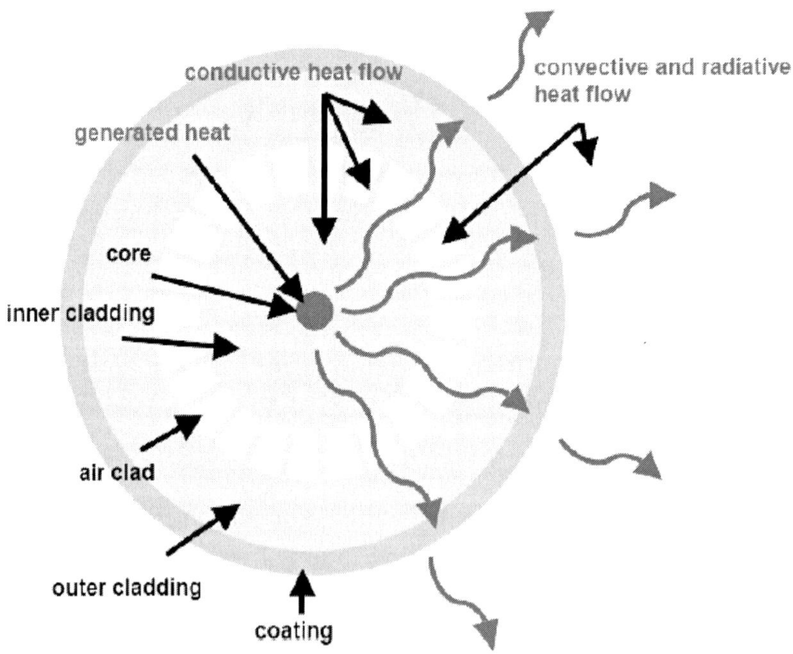

Figure 2. Heat flow mechanisms in PCF laser [9] .

We calculate the average temperature T_{av} as

$$T_{av} = \frac{\int_0^a T_1(r)\mathrm{d}r + \int_a^{b_1} T_2(r)\mathrm{d}r + \int_{b_1}^{b_2} T_3(r)\mathrm{d}r + \int_{b_2}^{b} T_4(r)\mathrm{d}r}{\int_0^b \mathrm{d}r} \tag{1}$$

where

$$\int_0^b \mathrm{d}r = b \tag{2}$$

$$\int_0^a T_1(r)\mathrm{d}r = T_0 a - \frac{Q(x)\times a^3}{12k_1} \tag{3}$$

$$\int_a^{b_1} T_2(r)\mathrm{d}r = T_1(a)(b_1 - a) + A(x)\times\left[b_1 \ln\left(\frac{b_1}{a}\right) - b_1 + a\right] \tag{4}$$

$$\int_{b_1}^{b_2} T_3(r)\mathrm{d}r = T_2(b_1)(b_2 - b_1) + B(x)\times\left[b_2 \ln\left(\frac{b_2}{b_1}\right) b_2 + b_1\right] \tag{5}$$

$$\int_{b_2}^{b} T_4(r)\mathrm{d}r = T_3(b_2)(b - b_2) + C(x)\times\left[b \ln\left(\frac{b}{b_2}\right) - b + b_2\right] \tag{6}$$

where the temperature expressions $T_1(r)$, $T_2(r)$, $T_3(r)$ and $T_4(r)$ are given by [8] .

And

$$A(x) = -\frac{Q(x)a^2}{2k_1} \tag{7}$$

$$B(x) = \left[-\frac{Q(x)a^2}{2k_2} + \frac{\varepsilon\sigma b_1}{k_2}\left(T_c^4 - T_2(b_1)^4\right)\right] \tag{8}$$

$$C(x)=\left[\frac{k_2}{k_1}\left(-\frac{Q(x)a^2}{2k_2}+\frac{\varepsilon\sigma b_1}{k_2}\left(T_c^4-T_2(b_1)^4\right)\right)-\frac{\varepsilon\sigma b_2}{k_1}\left(T_c^4-T_3(b_2)^4\right)\right] \tag{9}$$

3. STRESS DISTRIBUTIONS

The length of the optical fibers is much greater than a typical fiber outside radius (b), we can invoke the plane- strain approximation [10] in which the z strain everywhere. The radial, tangential, and z stresses $\sigma_r(r)$, $\sigma_\varphi(r)$, and $\sigma_z(r)$ can be found from [11] .

$$\sigma_r(r)=\frac{\alpha E}{(1-\vartheta)}\left[\frac{1}{b^2}\int_0^b T(r)r\mathrm{d}r-\frac{1}{r^2}\int_0^r T(r)r\mathrm{d}r\right] \tag{10}$$

$$\sigma_\varphi(r)=\frac{\alpha E}{(1-\vartheta)}\left[\frac{1}{b^2}\int_0^b T(r)r\mathrm{d}r-\frac{1}{r^2}\int_0^r T(r)r\mathrm{d}r-T(r)\right] \tag{11}$$

in the case where the fiber end faces are free of traction

$$\sigma_z(r)=\sigma_r(r)+\sigma_\varphi(r) \tag{12}$$

where α, E and ν are thermal expansion coefficient, Young's modulus, and Poisson's ratio, respectively. So- lution of (10)-(12), using the expressions previously derived for the temperatures in Regions I, II, III, and VI [8] , results in the following final expressions for the stresses in Regions I, II, III, and IV:

$$\sigma_r^I(r)=\frac{\alpha E}{(1-\vartheta)}\left[\frac{1}{b^2}\left(\alpha(x)+\beta(x)+\gamma(x)+\delta(x)\right)-\frac{1}{r^2}\left(\alpha'(x)\right)\right] \tag{13}$$

$$\sigma_r^{II}(r)=\frac{\alpha E}{(1-\vartheta)}\left[\frac{1}{b^2}\left(\alpha(x)+\beta(x)+\gamma(x)+\delta(x)\right)-\frac{1}{r^2}\left(\alpha(x)+\beta'(x)\right)\right] \tag{14}$$

$$\sigma_r^{III}(r)=\frac{\alpha E}{(1-\vartheta)}\left[\frac{1}{b^2}\big(\alpha(x)+\beta(x)+\gamma(x)+\delta(x)\big)-\frac{1}{r^2}\big(\alpha(x)+\beta(x)+\gamma'(x)\big)\right] \tag{15}$$

$$\sigma_r^{IV}(r)=\frac{\alpha E}{(1-\vartheta)}\left[\frac{1}{b^2}\big(\alpha(x)+\beta(x)+\gamma(x)+\delta(x)\big)-\frac{1}{r^2}\big(\alpha(x)+\beta(x)+\gamma(x)+\delta'(x)\big)\right] \tag{16}$$

$$\sigma_\varphi^{I}(r)=\frac{\alpha E}{(1-\vartheta)}\left[\frac{1}{b^2}\big(\alpha(x)+\beta(x)+\gamma(x)+\delta(x)\big)+\frac{1}{r^2}\big(\alpha'(x)\big)-T_1(r)\right] \tag{17}$$

$$\sigma_\varphi^{II}(r)=\frac{\alpha E}{(1-\vartheta)}\left[\frac{1}{b^2}\big(\alpha(x)+\beta(x)+\gamma(x)+\delta(x)\big)+\frac{1}{r^2}\big(\alpha(x)+\beta'(x)\big)-T_2(r)\right] \tag{18}$$

$$\sigma_\varphi^{III}(r)=\frac{\alpha E}{(1-\vartheta)}\left[\frac{1}{b^2}\big(\alpha(x)+\beta(x)+\gamma(x)+\delta(x)\big)+\frac{1}{r^2}\big(\alpha(x)+\beta(x)+\gamma'(x)\big)-T_3(r)\right] \tag{19}$$

$$\sigma_\varphi^{IV}(r)=\frac{\alpha E}{(1-\vartheta)}\left[\frac{1}{b^2}\big(\alpha(x)+\beta(x)+\gamma(x)+\delta(x)\big)+\frac{1}{r^2}\big(\alpha(x)+\beta(x)+\gamma(x)+\delta'(x)\big)-T_4(r)\right] \tag{20}$$

$$\sigma_z^{I}(r)=\frac{\alpha E}{(1-\vartheta)}\left[\frac{2}{b^2}\big(\alpha(x)+\beta(x)+\gamma(x)+\delta(x)\big)-T_1(r)\right] \tag{21}$$

$$\sigma_z^{II}(r)=\frac{\alpha E}{(1-\vartheta)}\left[\frac{2}{b^2}\big(\alpha(x)+\beta(x)+\gamma(x)+\delta(x)\big)-T_2(r)\right] \tag{22}$$

$$\sigma_z^{III}(r)=\frac{\alpha E}{(1-\vartheta)}\left[\frac{2}{b^2}\big(\alpha(x)+\beta(x)+\gamma(x)+\delta(x)\big)-T_3(r)\right] \tag{23}$$

$$\sigma_Z^{IV}(r)=\frac{\alpha E}{(1-\vartheta)}\left[\frac{2}{b^2}\big(\alpha(x)+\beta(x)+\gamma(x)+\delta(x)\big)-T_4(r)\right] \tag{24}$$

where.

$$\alpha(x)=\int_0^a T_1(r)r\mathrm{d}r=T_0\frac{a^2}{2}-\frac{Q(x)a^4}{16k_1} \tag{25}$$

$$\beta(x)=\int_a^{b_1} T_2(r)r\mathrm{d}r=T_1(a)\left(\frac{b_1^2-a^2}{2}\right)+A(x)\left[\frac{b_1^2}{2}\ln\left(\frac{b_1}{a}\right)-\frac{b_1^2}{4}+\frac{a^2}{4}\right] \tag{26}$$

$$\gamma(x)=\int_{b_1}^{b_2} T_3(r)r\mathrm{d}r=T_2(b_1)\left(\frac{b_2^2-b_1^2}{2}\right)+B(x)\left[\frac{b_2^2}{2}\ln\left(\frac{b_2}{b_1}\right)-\frac{b_2^2}{4}+\frac{b_1^2}{4}\right] \tag{27}$$

$$\delta(x)=\int_{b_2}^{b} T_4(r)r\mathrm{d}r=T_3(b_2)\left(\frac{b^2-b_2^2}{2}\right)+C(x)\left[\frac{b^2}{2}\ln\left(\frac{b}{b_2}\right)-\frac{b^2}{4}+\frac{b_2^2}{4}\right] \tag{28}$$

$$\alpha'(x)=\int_0^r T_1(r)r\mathrm{d}r=T_0\frac{r^2}{2}-\frac{Q(x)r^4}{16k_1} \tag{29}$$

$$\beta'(x)=\int_a^r T_2(r)r\mathrm{d}r=T_1(a)\left(\frac{r^2-a^2}{2}\right)+A(x)\left[\frac{r^2}{2}\ln\left(\frac{r}{a}\right)-\frac{r^2}{4}+\frac{a^2}{4}\right] \tag{30}$$

$$\gamma'(x)=\int_{b_1}^r T_3(r)r\mathrm{d}r=T_2(b_1)\left(\frac{r^2-b_1^2}{2}\right)+B(x)\left[\frac{r^2}{2}\ln\left(\frac{r}{b_1}\right)-\frac{r^2}{4}+\frac{b_1^2}{4}\right] \tag{31}$$

$$\delta'(x)=\int_{b_2}^r T_4(r)r\mathrm{d}r=T_3(b_2)\left(\frac{r^2-b_2^2}{2}\right)+C(x)\left[\frac{r^2}{2}\ln\left(\frac{r}{b_2}\right)-\frac{r^2}{4}+\frac{b_2^2}{4}\right] \tag{32}$$

It can be shown that (13)-(24) satisfy the boundary conditions $\sigma_r^{IV}(r=b)=0$, $\sigma_r^{I}(r=0)=\sigma_r^{I}(r=0)$, $\sigma_\varphi^{IV}(r=b)=\sigma_z^{IV}(r=b)$, and the continuity conditions $\sigma_r^{I}(r=a)=\sigma_r^{II}(r=a)$, $\sigma_\varphi^{I}(r=a)=\sigma_\varphi^{II}(r=a)$, $\sigma_z^{I}(r=a)=\sigma_z^{II}(r=a)$, $\sigma_r^{II}(r=b_1)=\sigma_r^{III}(r=b_1)$, $\sigma_\varphi^{II}(r=b_1)=\sigma_\varphi^{III}(r=b_1)$, $\sigma_z^{II}(r=b_1)=\sigma_z^{III}(r=b_1)$, $\sigma_r^{III}(r=b_2)=\sigma_r^{IV}(r=b_2)$, $\sigma_\varphi^{III}(r=b_2)=\sigma_\varphi^{IV}(r=b_2)$, and $\sigma_z^{III}(r=b_2)=\sigma_z^{IV}(r=b_2)$. Equations (13)-(24)

will be used in Section IV under to calculate the radially varying index of refraction in different pump schemes, due to the stress-optic effect. Note that the equations as derived above for the fiber stresses could also have been obtained by use of the deviation of the temperature distribution from the average and the Airy stress potential [11] .

4. INDEX OF REFRACTIONS

Using the expressions for the radial temperature distributions, reported [8] and for the stress distributions (13)- (24), we tray now to make calculation of the radially varying radial and tangential index of refraction distribu- tions. For comparison in different pump schemes, we prefer to cast the calculation of the induced change in the index of refraction in terms of material stresses, rather than strains. We begin by writing the changes in the in- dices of refraction:

$$\Delta n_{r,\varphi}^{I,II,III,IV}(r) = \Delta n_{r,\varphi}^{I,II,III,IV} - n_0 = \Delta n_{\beta}^{I,II,III,IV}(r) + \Delta n_{ST,r,\varphi}^{I,II,III,IV}(r) \tag{33}$$

where

$n_0 = 1.45$ Linear index of refraction;

$\Delta n_{\beta}^{I,II,III,IV}(r)$ Change in index due to change in index with temperature $\left(\beta = \frac{\mathrm{d}n}{\mathrm{d}T}\right)$;

$\Delta n_{r,\varphi}^{I,II,III,IV}(r)$ Stress-induced index changes for the r and φ components of the electric field.

As before, the roman numerals refer to Regions I, II, III and IV. Here, we will calculate $\Delta n_{\beta}^{I,II,III,IV}(r)$ using the following equation:

$$\Delta n_{\beta}^{I,II,III,IV}(r)=\beta\left(T_{1,2,3,4}(r)-T_c\right) \tag{34}$$

For Regions I, II, III and IV, becomes

$$\Delta n_{\beta}^{I}(r)=\beta\left(T_0-\frac{Q(x)\times r^2}{4k_1}-T_c\right) \tag{35}$$

And

$$\Delta n_{\beta}^{II}(r)=\beta\left(T_1(a)+A(x)\times\ln\left(\frac{r}{a}\right)-T_c\right) \tag{36}$$

And

$$\Delta n_{\beta}^{III}(r)=\beta\left(T_2(b_1)+B(x)\times\ln\left(\frac{r}{b_1}\right)-T_c\right) \tag{37}$$

And

$$\Delta n_{\beta}^{IV}(r)=\beta\left(T_3(b_2)+C(x)\times\ln\left(\frac{r}{b_2}\right)-T_c\right) \tag{38}$$

The calculation of the stress-induced index changes $\Delta n_{ST,r,\varphi}^{I,II,III,IV}(r)$ are treated in [10] .

Using

$$\Delta n_{ST,r}^{I}(r)=-\frac{n_0^3}{2}\left[B_{\perp}\left(\sigma_{\varphi}^{I}(r)+\sigma_{z}^{I}(r)\right)+B_{\mathrm{p}}\sigma_{r}^{I}(r)\right] \tag{39}$$

$$\Delta n_{ST,\varphi}^{I}(r)=-\frac{n_0^3}{2}\left[B_{\mathrm{p}}\sigma_{\varphi}^{I}(r)+B_{\perp}\left(\sigma_{z}^{I}(r)+\sigma_{r}^{I}(r)\right)\right] \tag{40}$$

$$\Delta n_{ST,r}^{II}(r)=-\frac{n_0^3}{2}\left[B_{\perp}\left(\sigma_{\varphi}^{II}(r)+\sigma_{z}^{II}(r)\right)+B_{\mathrm{p}}\sigma_{r}^{II}(r)\right] \tag{41}$$

$$\Delta n_{ST,\varphi}^{II}(r) = -\frac{n_0^3}{2}\left[B_\perp\left(\sigma_\varphi^I(r)+\sigma_z^I(r)\right)+B_p\sigma_r^I(r)\right] \tag{42}$$

$$\Delta n_{ST,r}^{III}(r) = -\frac{n_0^3}{2}\left[B_\perp\left(\sigma_\varphi^{III}(r)+\sigma_z^{III}(r)\right)+B_p\sigma_r^{III}(r)\right] \tag{43}$$

$$\Delta n_{ST,\varphi}^{III}(r) = -\frac{n_0^3}{2}\left[B_p\sigma_\varphi^{III}(r)+B_\perp\left(\sigma_z^{III}(r)+\sigma_r^{III}(r)\right)\right] \tag{44}$$

$$\Delta n_{ST,r}^{IV}(r) = -\frac{n_0^3}{2}\left[B_\perp\left(\sigma_\varphi^{IV}(r)+\sigma_z^{IV}(r)\right)+B_p\sigma_r^{IV}(r)\right] \tag{45}$$

$$\Delta n_{ST,\varphi}^{IV}(r) = -\frac{n_0^3}{2}\left[B_p\sigma_\varphi^{IV}(r)+B_\perp\left(\sigma_z^{IV}(r)+\sigma_r^{IV}(r)\right)\right] \tag{46}$$

where B_p and $B_\perp$ are the parallel and perpendicular stress-optic coefficients. Their values are $4.5\times10^{-8}\ \mathrm{cm^2/kg}$, $27.7\times10^{-8}\ \mathrm{cm^2/kg}$ respectively [11] . The numerical values of E and ν are $E = 73\ \mathrm{GPa} = 7.45\ \mathrm{kg/cm^2}$ and $\nu = 0.16$ [12] .

$$\Delta n_r^I(r) = \beta\left(T_0 - \frac{Q(x)\times r^2}{4k_1} - T_c\right) - \frac{n_0^3}{2}\left[B_\perp\left(\sigma_\varphi^I(r)+\sigma_z^I(r)\right)+B_p\sigma_r^I(r)\right] \tag{47}$$

$$\Delta n_\varphi^I(r) = \beta\left(T_0 - \frac{Q(x)\times r^2}{4k_1} - T_c\right) - \frac{n_0^3}{2}\left[B_p\sigma_\varphi^I(r)+B_\perp\left(\sigma_z^I(r)+\sigma_r^I(r)\right)\right] \tag{48}$$

$$\Delta n_r^{II}(r) = \beta\left(T_1(a) + A(x)\times\ln\left(\frac{r}{a}\right) - T_c\right) - \frac{n_0^3}{2}\left[B_\perp\left(\sigma_\varphi^{II}(r)+\sigma_z^{II}(r)\right)+B_p\sigma_r^{II}(r)\right] \tag{49}$$

$$\Delta n_\varphi^{II}(r) = \beta\left(T_1(a) + A(x)\times\ln\left(\frac{r}{a}\right) - T_c\right) - \frac{n_0^3}{2}\left[B_p\sigma_\varphi^{II}(r)+B_\perp\left(\sigma_r^{II}(r)+\sigma_z^{II}(r)\right)\right] \tag{50}$$

$$\Delta n_r^{III}(r)=\beta\left(T_2(b_1)+B(x)\times\ln\left(\frac{r}{b_1}\right)-T_c\right)-\frac{n_0^3}{2}\left[B_\perp\left(\sigma_\varphi^{III}(r)+\sigma_z^{III}(r)\right)+B_\text{p}\sigma_r^{III}(r)\right] \tag{51}$$

$$\Delta n_\varphi^{III}(r)=\beta\left(T_2(b_1)+B(x)\times\ln\left(\frac{r}{b_1}\right)-T_c\right)-\frac{n_0^3}{2}\left[B_\text{p}\sigma_\varphi^{III}(r)+B_\perp\left(\sigma_z^{III}(r)+\sigma_r^{III}(r)\right)\right] \tag{52}$$

$$\Delta n_r^{IV}(r)=\beta\left(T_3(b_2)+C(x)\times\ln\left(\frac{r}{b_2}\right)-T_c\right)-\frac{n_0^3}{2}\left[B_\perp\left(\sigma_\varphi^{IV}(r)+\sigma_z^{IV}(r)\right)+B_\text{p}\sigma_r^{IV}(r)\right] \tag{53}$$

$$\Delta n_\varphi^{IV}(r)=\beta\left(T_3(b_2)+C(x)\times\ln\left(\frac{r}{b_2}\right)-T_c\right)-\frac{n_0^3}{2}\left[B_\text{p}\sigma_\varphi^{IV}(r)+B_\perp\left(\sigma_z^{IV}(r)+\sigma_r^{IV}(r)\right)\right] \tag{54}$$

We can also define the brief ringence $\Delta n_B^{I,II,III,IV}(r)$, given by

$$\Delta n_B^{I,II,III,IV}(r)=\Delta n_r^{I,II,III,IV}(r)-\Delta n_\varphi^{I,II,III,IV}(r) \tag{55}$$

Which may be calculated by using, yielding

$$\Delta n_B^{I}(r)=\frac{n_0^3}{2}B\left(\sigma_r^{I}(r)-\sigma_\varphi^{I}(r)\right) \tag{56}$$

$$\Delta n_B^{II}(r)=\frac{n_0^3}{2}B\left(\sigma_r^{II}(r)-\sigma_\varphi^{II}(r)\right) \tag{57}$$

$$\Delta n_B^{III}(r)=\frac{n_0^3}{2}B\left(\sigma_r^{III}(r)-\sigma_\varphi^{III}(r)\right) \tag{58}$$

$$\Delta n_B^{IV}(r)=\frac{n_0^3}{2}B\left(\sigma_r^{IV}(r)-\sigma_\varphi^{IV}(r)\right) \tag{59}$$

Equations (56)-(59) show that the fiber birefringence depends only on the thermally induced stresses. Finally, by using (43)-(50) and substituting (13)-(20) for the radial, tangential, and z stress components, we arrive at the final equations describing the radially varying indices of refraction in a fiber.

Where $\Delta n_r^I(r)$ and $\Delta n_\varphi^I(r)$, apart from a constant term, vary quadratically with the fiber radius, in agree- ment with previously published work on bulk rod laser amplifiers [13] [14] . For Region II, however, it is noted that $\Delta n_r^{II}(r)$ and $\Delta n_\varphi^{II}(r)$ vary in a complicated way that involves both a logarithmic function and an inverse square.

5. DISCUSSIONS

For simplicity, we concentrated in this work on Yb^{3+}. The core region is surrounded by a circle inner cladding region with dimensions of radius 125 μm, the width of air-clad is 5 μm and which is in turn surrounded by a polymeric outer cladding region with outside diameter of about 300 μm.

5.1. Stress Effect

In Figures 3(a)-(e), we plotted the stress components radial, tangential and longitudinal as a function of the radial coordinate r in different pump schemes for a Yb^{3+} fiber with a pump of 200 W, b = 300 μm is the outer radius, convective coefficient h = 40.9 W/cm×K. The quantities $B_\perp$ and $B_\|$ are known in the optics and laser literature as the perpendicular and parallel stress-optic coefficients their values are found in Section 4, the thermal expansion coefficient is $\alpha = 0.51\times10^{-6}/\mathrm{k}$ [15] , we show no difference between (a) and (b) in side of the pumping (left side). Moreover, a small difference in other side (side right), we show the same thing for (c) and (d) figures. However, in Figure 3(e) the value of stress is smaller than the previous cases. The difference between the maximum and minimum values of stress components, in the four previous cases (a, b, c, d) is in the order 2×10^{-6} kg/m^2 but in bi-directional pump scheme Figure 3(e) the value is the order 1×10^{-6} kg/m^2. Therefore in bi-directional pump scheme the effects of the stress is negligible with regard to the others cases.

For more clarification, we plotted the stress components as a function of the radius in different pump schemes. In Figures 4(a)-(e), we observe that there is a rapid increase in air-clad region of the components radial $\sigma_r(r)$ and longitudinal $\sigma_z(r)$. However, concerning the tangential component $\sigma_\varphi(r)$, the increase is less rapid. We notice that the required boundary conditions are all satisfied. At r = b, the tangential $\sigma_\varphi(r)$ and longitudinal $\sigma_z(r)$ stresses are equal while the radial $\sigma_r(r)$ stress is zero along the fiber. Therefore, we anticipate that changes in the index of refraction due to the stress optic effect will be negligible in all cases of pumping and precisely in bi-directional pump scheme and thermally induced birefringence to be absent in fiber lasers.

5.2. Index of Refraction

Using (47)-(54), we plotted the tangential and the radial indices of refraction as a function of the radial coordinate r in different pump schemes, Figures 5(a)-(e), the pump power used is 200 W, the fiber outside radius is b = 300 μm, and the convection coefficient is h = 40.9 W/m×K, and $\beta=10^{-5}/\text{k}$ [12] . The curves for $\Delta n_r(r)$ and $\Delta n_\varphi(r)$ are essentially identical in r = 0 μm and in r = b, and the radially varying index is due to $\beta=\frac{dn}{dT}$ only, because $\sigma_r(r)$ and $\sigma_\varphi(r)$ are so small. The values of the change in index of refraction decrease along the fiber lasers in Figure 5(a) and Figure 5(b) from the order 1.2×10^{-3} to 0.1×10^{-3}. In Figure 5(c) and Figure 5(d) the values of change in index of refraction increase in the opposite order 0.1×10^{-3} to 1.2×10^{-3}.

Nevertheless, in Figure 5(e) the value is in order of 3×10^{-3} to 7×10^{-3}. As result, we notice in side of pumping there is a small difference of the change in index of refraction between the fiber core r = 0 and in the outer clad r = b, the four Figures 5(a)-(d). However, in Figure 5(e) the difference of the change in index of re- fraction between the fiber core r = 0 and in the outer clad r = b is small along the fiber.

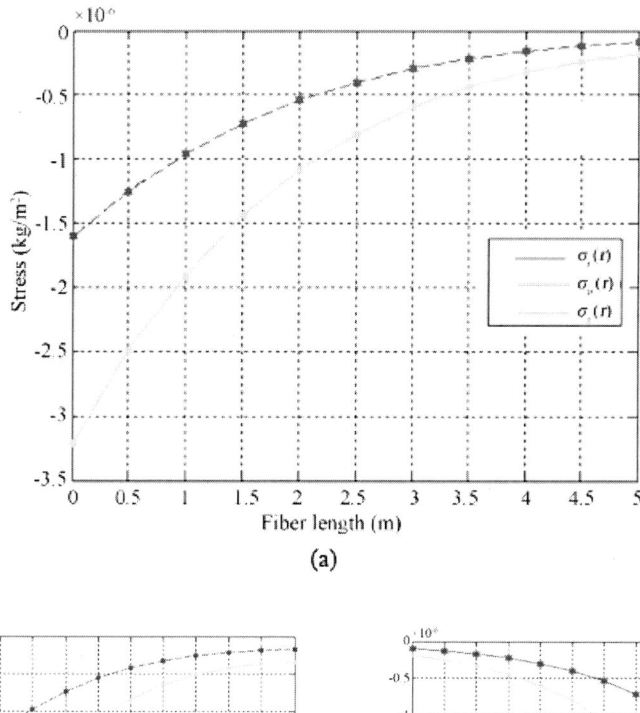

(a)

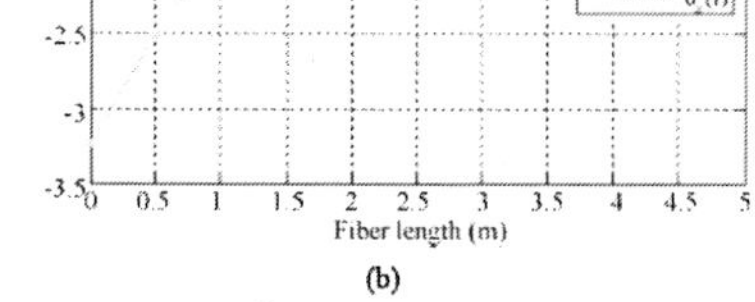

(b)

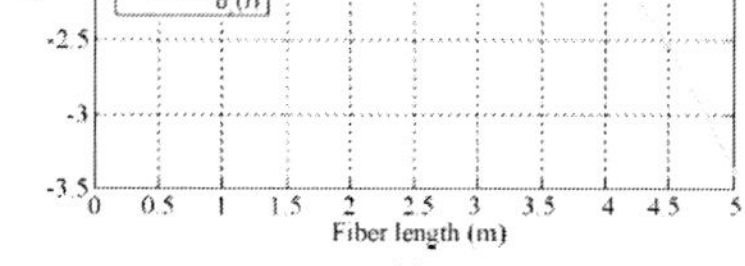

(c)

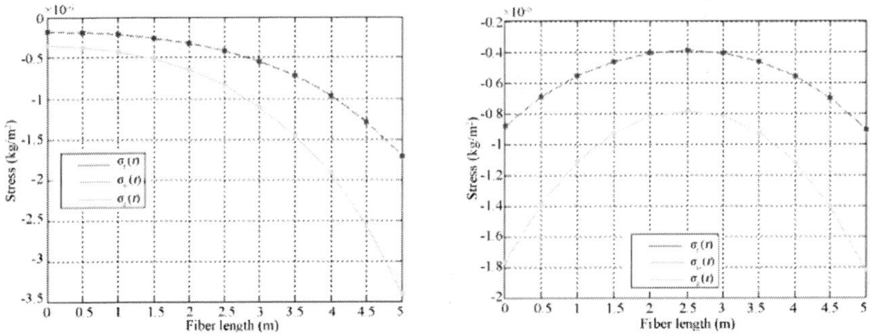

(d) (e)

Figure 3. Distribution of longitudinal stress in PCF at 200 W pump of 940 nm load calculated by FDM analysis. (a) Forward pump schemes; (b) Forward pump with a reflectivity of 98%; (c) Backward pump schemes; (d) Backward pump with a reflectivity of 98%; (e) Tow-end pump schemes.

6. SUMMARY

In this paper, we have investigated a comparison of stress and thermo-optic of photonic crystal fiber (PCFs) in different pump schemes. Using in these calculations a simple model of (PCFs) and finite differential method (FDM), we have revealed the temperature in the core of the fiber and by laws of heat transfer, we determined its value at the surface of the fiber and the stress value in the different regions of the fiber.

M. Abouricha, A. Boulezhar, M. El Mouden, M. Kriraa, N. Rouchdi In summary, regarding stress, thermo-optic and the change in index of refraction, their value does not have a great effect on the quality of the laser beam in different pump schemes, especially in the case of the bi-direc- tional pumping. Hence, after this investigation, we proved the architecture of laser cavity that was the most convenient in specific condition. The bi-directional pump scheme is the most suitable because it has less thermal effects than the other cases. However, the backward pump with reflection of 98% and the forward pump with reflection of 98% save more energy than the backward pump with reflection of 0%, the forward pump with ref- lection of 0% and bi-directional pump scheme.

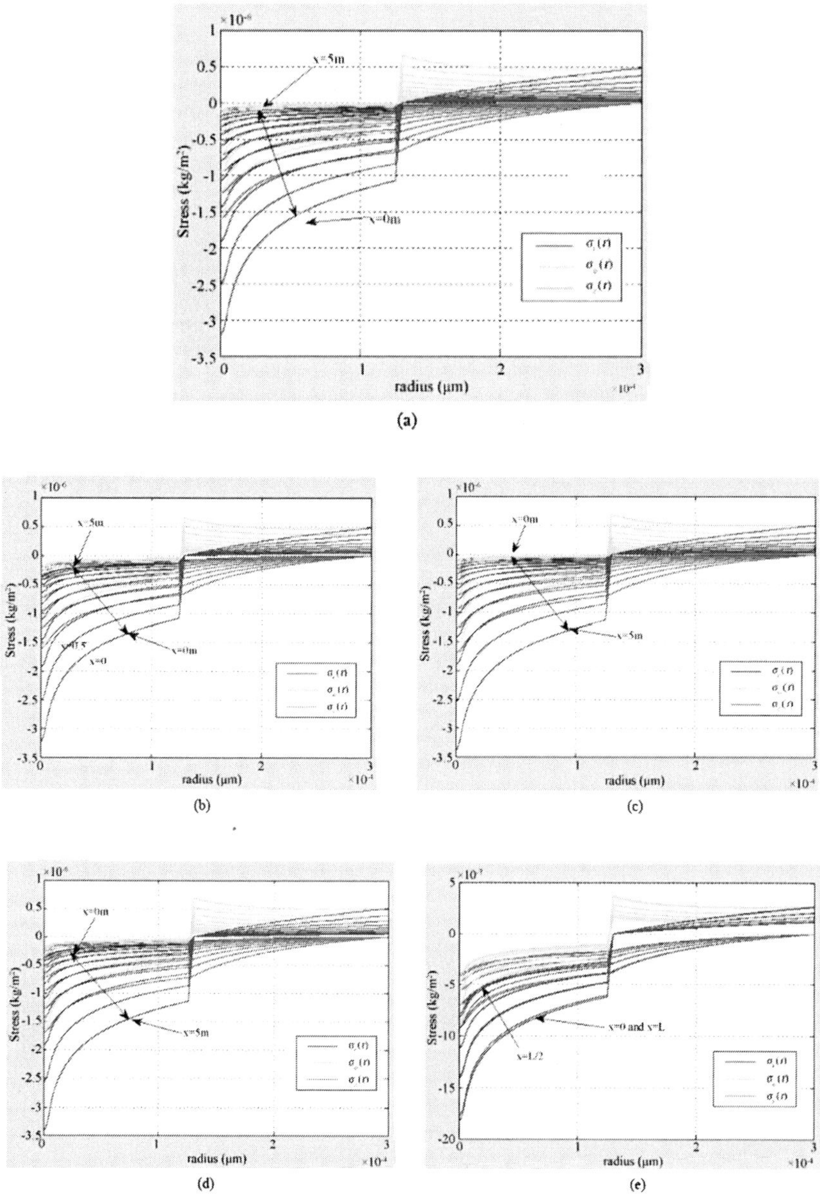

Figure 4. Distribution of radial stress in PCF at 200 W of 940 nm pump load calculated by FDM analysis. (a) Forward pump schemes; (b) Forward pump with a reflectivity of 98%; (c) Backward pump schemes; (d) Backward pump with a reflectivity of 98%; (e) Tow-end pump schemes.

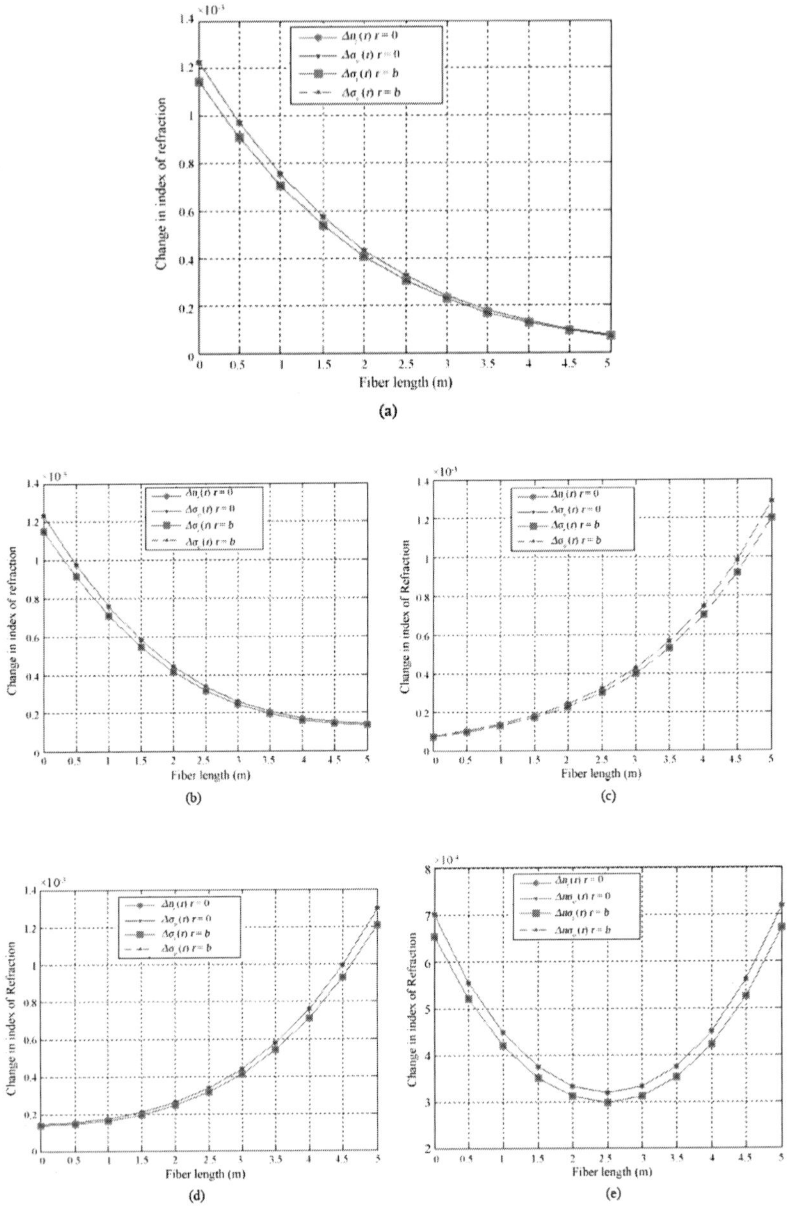

Figure 5. The change in index of refraction in PCF at 200 W pump load calculated by FDM analysis. (a) Forward pump schemes; (b) Forward pump with a reflectivity of 98%; (c) Backward pump schemes; (d) Backward pump with a reflectivity of 98%; (e) Tow-end pump schemes.

REFERENCES

1. Limpert, J., Liem, A., Zellmer, H. and Tunnerman, A. (2003) 500 W Continuous-Wave Fiber Laser with Excellent Beam Quality. Electronics Letters, 39, 645-647.

2. DiGiovanni, D.L. and Muendel, M.H. (1999) High Power Fiber Lasers and Amplifiers. Optics and Photonics News, 10, 26-30.

3. Platonov, N., Gapontsev, V.P., Shkurihin, O. and Zaitsev, I. (2003) 400 W Low-Noise Single-Mode CW Ytterbium Fiber La- ser with an Integrated Fiber Delivery. Conference on Lasers and Electro-Optics, Optical Society of America, Washington DC, 6 June 2003.

4. Muendel, M.H. (1999) High-Power Fiber Laser Studies at the Polaroid Corporation. Proceedings of the SPIE Conference High-Power Lasers, San Jose, 24 June 1998, Vol. 3264, 21-29.

5. Shang, L. (2011) Comparative Study of the Output Characteristics of Ytterbium-Doped Double-Clad Fiber Lasers with Different Pump Schemes. Optik, 122, 1899-1902.

6. Knight, J.C. and Russell, P.St.J. (2002) New Ways to Guide Light. Science, 296, 276-277.

7. Knight, J.C., Briks, T.A., Cregan, R.F., et al., (1998) Large Mode Area Photonic Crystal Fiber. Electronics Letters, 34, 1347-1349.

8. Abouricha, M., Boulezhar, A. and Habiballah, N. (2013) The Comparative Study of the Temperature Distribution of Fiber Laser with Different Pump Schemes. Open Journal of Metal, 3, 64-71.

9. Limpert, J., Schreiber, T., Liem, A., Nolte, S. and Zellmer, H. (2003) Thermo-Optical Properties of Air-Clad Photonic Crystal Fiber Lasers in High Power Operation. Optics Express, 11, 2982-2990.

10. Boley, B.A. and Weiner, J.H. (1985) Theory of Thermal Stresses. Kreiger, Melbourne.

11. Brown, D.C. and Hoffman, H.J., IEEE Member (2001) Thermal, Stress, and Thermo-Optic Effects in High Average Power Double-Clad Silica Fiber Lasers. IEEE Journal of Quantum Electronics, 37, 207-217.

12. Bass, M., Ed. (1995) Handbook of Optics (Sponsored by the Optical Society of America). Vol. II, Devices, Measurements, and Properties. McGraw-Hill, New York.

13. Eggleston, J.M., Kane, T.J., Kuhn, K., Unternahrer, J. and Byer, R.L. (1984) The Slab Geometry Laser—Part I: Theory. The IEEE Journal of Quantum Electronics, QE-20, 289-301.

14. Brown, D.C. (1998) Nonlinear Thermal Distortion in YAG Rod Amplifiers. IEEE Journal of Quantum Electronics, 34, 2383-2382.

15. Krupke, W.F., Shinn, M.D., Marion, J.E., Caird, J.A. and Stokowski, S.E. (1986) Spectroscopic, Optical, and Thermo- mechanical Properties of Neodymium- and Chromium-Doped Gadolinium Scandium Gallium Garnet. Journal of the Optical Society of America B, 3, 102, 113.

12

COMBINATORIAL FREQUENCY GENERATION IN QUASI-PERIODIC STACKS OF NONLINEAR DIELECTRIC LAYERS

Oksana Shramkova and Alexander Schuchinsky *

School of Electronics, Electrical Engineering & Computer Science,
Queen's University Belfast, Queen's Road, Queen's Island,
Belfast BT3 9DT, UK

ABSTRACT

Three-wave mixing in quasi-periodic structures (QPSs) composed of nonlinear anisotropic dielectric layers, stacked in Fibonacci and Thue-Morse sequences, has been explored at illumination by a pair of pump waves with dissimilar frequencies and incidence angles. A new formulation of the nonlinear scattering problem has enabled the QPS analysis as a perturbed periodic structure with defects. The obtained solutions have revealed the effects of stack composition and constituent layer parameters, including losses, on the properties of combinatorial frequency generation (CFG). The CFG features illustrated by the simulation results are discussed. It is demonstrated that quasi-periodic stacks can achieve a higher efficiency of CFG than regular periodic multilayers.

KEYWORDS

quasi-periodic structure; nonlinear dielectric; nonlinear scattering; three-wave mixing; combinatorial frequency generation; Fibonacci stack; Thue-Morse sequence

1. INTRODUCTION

The recent theoretical and experimental studies of nonlinear metamaterials (MM) and photonic crystals (PhC) have demonstrated potential of their applications in functional devices operating in diverse frequency ranges, from microwaves to optics. Traditionally, artificial media have periodic order with regular spectral characteristics. In contrast, long-range order of quasi-periodic structures (QPSs) and quasi-crystals provides additional degrees of freedom and enables new unusual properties, which recently attracted considerable interest [1].

QPSs with high order symmetry lie between periodic crystals and random amorphous solids. Since the first realization of Fibonacci and Thue-Morse type superlattices [2], the physical properties of 1-D QPSs have been extensively studied, both experimentally and theoretically, and it has been shown that QPSs can exhibit unique features unattainable in periodic arrangements. In particular, it was found that QPSs create localization of electronic states and their collective response differs from that of their constituents. While the linear wave phenomena in quasi-periodic media were extensively explored during the last few decades [3,4,5,6,7,8,9,10,11,12,13,14, 15,16,17,18,19,20,21,22,23,24], nonlinear QPSs have been studied only recently [25,26,27,28,29,30,31,32,33,34,35].

The main distinctive features of nonlinear artificial media are usually associated with their ability to simultaneously control the dispersion and nonlinear interactions by altering both the geometry of constituent particles and their arrangements in ensembles. Both periodic and quasi-periodic nonlinear structures have been studied primarily with the aim of improving harmonic generation efficiency. QPSs with additional degrees of freedom due to the higher order group symmetry are particularly attractive for creating higher density of states that can facilitate harmonic generation [36,37,38]. It has been observed in [25] that the efficiency of second-harmonic generation (SHG) in QPS is higher than in random medium, but still remains lower than in regular periodic structures. On the other hand, an eight-fold increase of the third-harmonic generation (THG) efficiency has been attained in the two-step cascaded process in a QPS as compared with a similar periodic multilayer [26].

Combinatorial frequency generation (CFG) by mixing pump waves of different tones enables more versatile spectrum control. However, the dominant three- and four-wave mixing processes are usually demanding to not only the nonlinearity itself but also the linear dispersion and field distribution, which primarily depend on the medium intrinsic properties and composition. Therefore, QPSs with their rich spectral properties may provide favourable conditions for frequency mixing and CFG. To realise this potential, it is important to explore in detail the mechanisms and features of CFG in QPSs.

In this paper, a generic approach to the analysis of CFG by layered QPSs, illuminated by two obliquely incident pump waves, has been developed and applied to the study of the properties of frequency mixing products. The problem statement and solution of the boundary value problem for the Fibonacci and Thue-Morse type QPSs are outlined in Section 2 and elaborated in Appendices. The features of CFG by the QPSs composed of nonlinear anisotropic dielectric multilayers are illustrated by the results of numerical simulations and discussed in Section 3. The main findings are summarised in Conclusion.

2. PROBLEM STATEMENT AND METHODOLOGY

QPSs shown in Figure 1 are composed of stacked weakly nonlinear anisotropic dielectric layers of two kinds (labelled *A*and *B*) with thicknesses d_A and d_B, respectively. Stacks of thicknesses L_q are surrounded by a linear homogeneous medium with the relative permittivity ε_a at $z \le 0$ and $z \ge L_q$. The constituent dielectric layers with quadratic nonlinearity and 6 mm class of anisotropy are described by the tensors of relative linear permittivity $\hat{\varepsilon}_j = (\varepsilon_{xx}, \varepsilon_{xx}, \varepsilon_{zz})_j$ and nonlinear susceptibility $\hat{\chi}_j$ where $j = A, B$:

$$\hat{\chi}_j = \begin{pmatrix} 0 & 0 & 0 & 0 & \chi_{xxz} & 0 \\ 0 & 0 & 0 & \chi_{xxz} & 0 & 0 \\ \chi_{zxx} & \chi_{zxx} & \chi_{zzz} & 0 & 0 & 0 \end{pmatrix}_j \quad (1)$$

The stacked layers are arranged in accordance with the following quasi-periodic sequences:

a) Fibonacci type QPS (Figure 1a) of order $q \ge 2$ is defined recursively by the recurrence relation: $S_q = \{S_{q-1} \cup S_{q-2}\}$, where $S_0 = \{B\}$, $S_1 = \{A\}$; the number of layers in a stack S_q equals Fibonacci number Φ_{q+1} and the corresponding stack thickness is $L_q = L_{q-1} + L_{q-2}$ with $L_0 = d_B$; $L_1 = d_A$;

b) Thue-Morse type QPS (Figure 1b) of order $q \ge 1$ is defined recursively by the recurrence relations: $Q_q = \{Q_{q-1} \cup Q'_{q-1}\}$ and $Q'_q = \{Q'_{q-1} \cup Q_{q-1}\}$, where Q_q and Q'_q are complementary to each other and $Q_0 = \{A\}$, $Q'_0 = \{B\}$; the stacks Q_q andQ'_q contain 2^q layers and have thicknesses $L_q = L'_q = (d_A + d_B)^q$ at $q > 0$ and $L_0 = d_A$, $L'_0 = d_B$.

The stacks are illuminated by two pump plane waves of frequencies ω_1 and ω_2 incident at angles Θ_{i1} and Θ_{i2}, respectively. Since the layers are assumed isotropic in the *x*-*y* plane, the TE and TM polarised waves with the fields independent of the *y*-coordinate ($\partial/\partial_y = 0$) can be analysed separately. Only TM waves with the field components E_x, E_z, $H_y$are presented in the paper. The TE waves are treated similarly and even somewhat simpler, because the nonlinear susceptibility given by Equation (1) is isotropic for TE polarised waves.

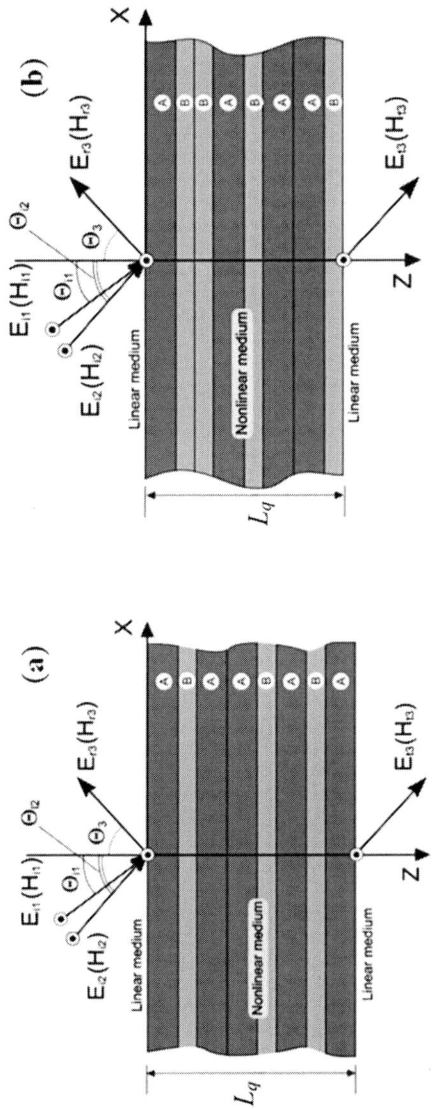

Figure 1. Geometry of the problem: (**a**) Fibonacci type QPS; (**b**) Thue-Morse type QPS.

To evaluate the characteristics of TM waves of combinatorial frequencies generated by a nonlinear QPS, the fields of both pump waves and their mixing products inside the stack are to be determined first. Assuming that nonlinearity of the constituent layers is weak and the three-wave mixing process is dominant, the nonlinear products can be obtained by the harmonic balance method. At combinatorial frequency $\omega_3 = \omega_1 + \omega_2$ the TM wave fields in each layer are described by non-homogeneous Helmholtz equation:

$$\frac{\partial^2 H_{yj}(\omega_3)}{\varepsilon_{xxj}\partial z^2} + \left(k_3^2 - \frac{k_{x3}^2}{\varepsilon_{zzj}}\right) H_{yj}(\omega_3) = 4\pi k_3 \left[2\frac{\partial}{\partial x}\left(\frac{\chi_{zxxj}}{\varepsilon_{zzj}} E_{xj}(\omega_1) E_{xj}(\omega_2) + \frac{\chi_{zzzj}}{\varepsilon_{zzj}} E_{zj}(\omega_1) E_{zj}(\omega_2)\right) - \frac{\chi_{xxzj}}{\varepsilon_{xxj}} \frac{\partial}{\partial z}\left(E_{xj}(\omega_1) E_{zj}(\omega_2) + E_{xj}(\omega_2) E_{zj}(\omega_1)\right)\right] \tag{2}$$

where $E_{xj}(\omega_{1,2})$ and $E_{zj}(\omega_{1,2})$ are the electric field components of pump waves in a layer of type $j = A, B$; $k_3 = \omega_3/c$ and $k_{x3} = k_3\sqrt{\varepsilon_a}\sin\Theta_3$; angle Θ_3 defines the emission direction from the stacks at frequency ω_3, which is determined by the phase synchronism condition in the three-wave mixing process [39]:

$$k_{x3} = k_{x1} + k_{x2} \tag{3}$$

where $k_{x1,x2} = k_{1,2}\sqrt{\varepsilon_a}\sin\Theta_{i1,i2}$, $k_{1,2} = \omega_{1,2}/c$

The solution of Equation (2) for $H_{yj}(\omega_3)$ can be represented as a superposition of six plane waves with longitudinal propagation constants $\pm k_{zj}^{\pm}$ and $\pm k_{zj}^{(3)}$.

$$H_{yj}^{(n)}(\omega_3, x, z) = \left(\mathcal{B}_j^{n+}(\omega_3) e^{ik_{zj}^{(3)} z} + \mathcal{B}_j^{n-}(\omega_3) e^{-ik_{zj}^{(3)} z} + D_{1j}^{n+} e^{ik_{zj}^{+} z} + D_{2j}^{n+} e^{-ik_{zj}^{+} z} + D_{1j}^{n-} e^{ik_{zj}^{-} z} + D_{2j}^{n-} e^{-ik_{zj}^{-} z}\right) e^{-i\omega_3 t + ik_{x3} x} \tag{4}$$

where $k_{zj}^{\pm} = k_{zj}^{(1)} \pm k_{zj}^{(2)}$, $k_{zj}^{(p)} = \sqrt{\varepsilon_{xxj}\left(k_p^2 - k_{xp}^2/\varepsilon_{zzj}\right)}$, p = 1,2,3 and superscript n is a serial number of a primitive cell, which identifies the position of a jth type layer in the stack. $D_{1j,2j}^{n\pm}$ are amplitudes of the waves of frequency ω3, generated inside this layer, which depend on amplitudes $\mathcal{B}_j^{n\pm}(\omega_{1,2})$ of the pump waves refracted into this layer [40]. $\mathcal{B}_j^{n\pm}(\omega_3)$ are amplitudes of the waves of frequency ω3 generated outside the layer and refracted into it.

To calculate the amplitude coefficients in Equation (4), it is necessary first to find the pump wave amplitudes in each constituent layer. In periodic stacks comprised of alternating binary layers this is accomplished by the transfer matrix method (TMM) [41,42,43,44] as the unit cell layout and serial number n fully identify the layer type (A or B) and position in the stack. This approach can be extended to QPS because the stack composition is defined by the deterministic rules. However, evaluation of each layer position and type in a specific sequence is usually a challenging task.

Examination of Fibonacci and Thue-Morse QPSs shows that they contain only two types of primitive cells. Indeed, Fibonacci QPS, for example, S_6 = {*AB AAB AB AAB AAB*} is comprised of "regular" cells {*AB*} and "defective" cells {*AAB*} with an additional A layer which forms a layer A doublet of thickness $d'_A = 2d_A$. Similarly, Thue-Morse QPS, for example, Q_4= {*AB BA BA AB BA AB AB BA*} is composed of the same "regular" cells {*AB*} and "defective" cells {*BA*} with inverse order of layers but the same cell thickness. Thus once the cell serial number n, and layout and position of the regular and defective cells are known, the conventional TMM can be adapted to obtain amplitude coefficients $D_{1j,2j}^{n\pm}$ and $\mathcal{B}_j^{n\pm}(\omega)$ in Equation (4) for Fibonacci and Thue-Morse QPSs as detailed in Appendix 1.

Once the fields of combinatorial frequencies ω3 in individual layers are determined, amplitudes $F_r(\omega_3)$ and $F_t(\omega_3)$ of the field emitted from stacks toward the reverse and forward directions of the z-axis, respectively, can be evaluated by the modified TMM [40] as described in Appendix 2. The obtained analytical solutions provide an insight in the fundamental mechanisms of nonlinear scattering and frequency mixing by the QPSs illuminated by a pair of obliquely incident plane waves. The features of CFG in Fibonacci and Thue-Morse QPSs are further illustrated by examples of numerical simulations and discussed in the next section.

3. SIMULATION RESULTS AND DISCUSSION

To elucidate the CFG mechanisms in QPSs, the nonlinear scattering characteristics of TM waves by the stacks of Fibonacci and Thue-Morse sequenced binary anisotropic layers have been analysed numerically. Illustrative examples of the simulation results for the specific QPS arrangements are discussed below in comparison with the corresponding periodic stacks of the same thicknesses.

It is assumed that the stacks are surrounded by air with the relative permittivity $\varepsilon_a = 1$ and are illuminated by a pair of pump plane waves of frequencies $\omega_{1,2}$ incident at angles $\Theta_{i1,i2}$ as shown in Figure 1. The QPS constituent layers are anisotropic dielectric films of CdS and ZnO with the following values of the tensor relative linear permittivity $\hat{\varepsilon}_{A,B}$ and nonlinear susceptibility $\bar{\chi}_{A,B}$ (the $\bar{\chi}_{A,B}$ units [m/V] are omitted below for brevity) [45]:

Layer A (CdS): $\varepsilon_{xxA} = 5.382$, $\varepsilon_{zzA} = 5.457$, $\chi_{xxzA} = 2.1 \times 10^{-7}$, $\chi_{zxxA} = 1.92 \times 10^{-7}$, $\chi_{zzzA} = 3.78 \times 10^{-7}$,

Layer B (ZnO) $\varepsilon_{xxB} = 1.4$, $\varepsilon_{zzB} = 2.6$, $\chi_{xxzB} = 2.82 \times 10^{-8}$, $\chi_{zxxB} = 2.58 \times 10^{-8}$, $\chi_{zzzB} = 8.58 \times 10^{-8}$.

The layer thicknesses d_A and d_B are related by the golden ratio: $\tau = d_A/d_B = \left(1+\sqrt{5}\right)/2$.

3.1. Spectral Efficiency of Frequency Mixing

The earlier studies of photonic crystals and periodic stacks of dielectric layers have suggested that the SHG and THG efficiency may be higher at the bandgap edges. However, the concurrent increase of losses considerably reduces the gain. Alternatively, CFG efficiency proved to be higher not at the band edges but in the transparency bands, especially at Wolf-Bragg resonances of the stacks [40]. Additionally, CFG involving two pump waves of different frequencies ω_1 and ω_2, incident at dissimilar angles, offers extra degrees of freedom in controlling the frequency mixing process and spectrum.

The frequency mixing efficiency in the stacks of nonlinear layers essentially depends on the phase coherence, which is controlled by linear dispersion of both pump waves and their nonlinear products. Additionally, magnitudes of interacting pump waves are significantly influenced by the stack composition and linear reflectance $R(\omega)$, variable with the incidence angle. Therefore rich spectral content of QPS has potential for assisting CFG enhancement.

To illustrate the effect of the stack composition, $|R(\omega)|^2$ of TM waves incident at angles $\Theta_i = 30°$ and $\Theta_i = 45°$ are displayed in Figure 2 for Fibonacci S_8 (34 layers, $d_B = 12$ μm) and Thue-Morse Q_5 (32 layers, $d_B = 13$ μm) QPSs in comparison with the periodic stack of alternating layers (16 unit cells or 32 layers, $d_B = 13$ μm). All three stacks have the same thickness, and equal number of A- and B-type layers in the periodic and Thue-Morse stacks, while the Fibonacci QPS contains more layers of smaller thicknesses. Considerable differences in $|R(\omega)|$ of the periodic stack and QPSs are evident in Figure 2, especially at higher frequencies. This implies that the frequency mixing efficiency may significantly vary due to disparity of the pump wave magnitudes refracted into the stack. Bandgaps, corresponding to $|R(\omega)| \approx 1$, exist in both periodic structures and QPSs, albeit more intricate spectra of QPSs provide more flexible conditions for CFG as discussed below.

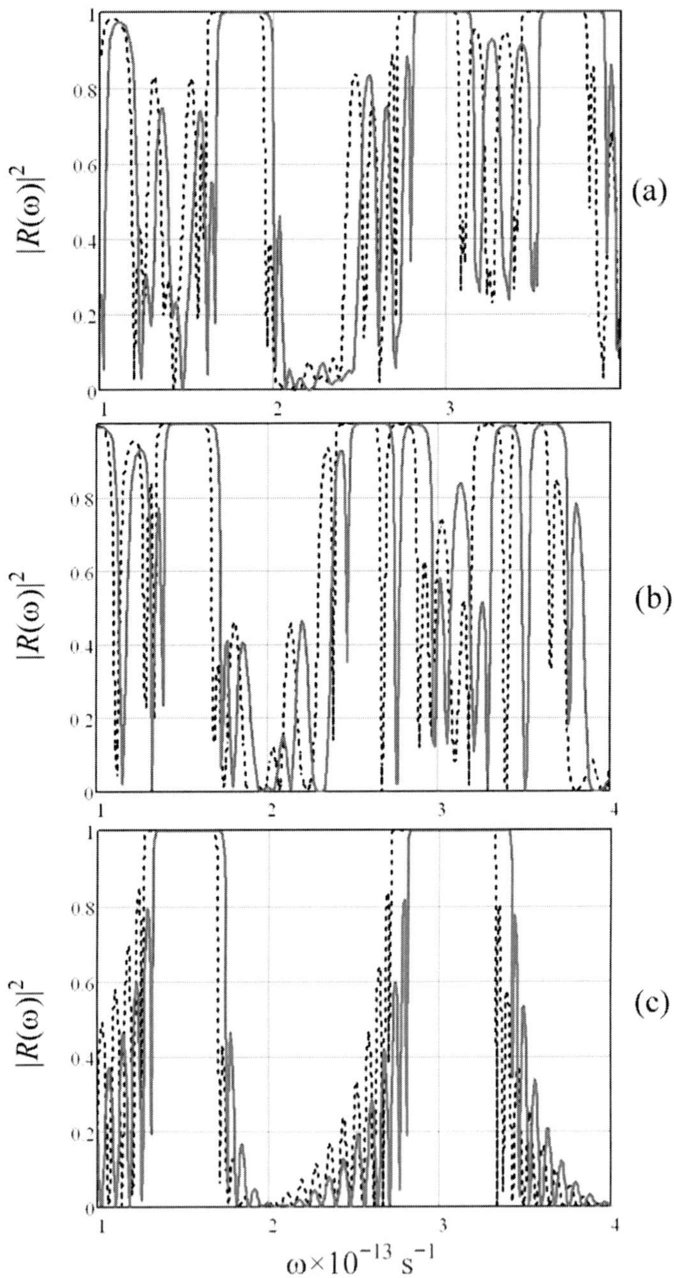

Figure 2. Reflectance of TM wave incident at Θ_i = 30° (dashed black lines) and Θ_i = 45° (solid red lines) on (**a**) Fibonacci S_8 and (**b**) Thue-Morse Q_5 QPSs, and (**c**) periodic stack with 16 unit cells.

To assess the effect of the pump wave reflectance on CFG, magnitudes $|F_{r,t}(\omega_3)|$ of the waves of combinatorial frequency ω_3 emitted from the stacks have been simulated at fixed frequencies ω_2 of a pump wave incident at $\Theta_{i2} = 45°$, and swept frequency ω_1 of the other pump wave incident at $\Theta_{i1} = 30°$. Frequencies ω_2 have been chosen individually for each stack configuration so that $|R(\omega_2)|^2$ be at the minima close to the transparency band edges. Since QPS bands are not clearly defined as in periodic stacks, we refer only to the reflectance minima, which depend on the constituent layer parameters and incidence angles of pump waves as illustrated by Figure 2.

Inspection of Figure 2 and Figure 3 demonstrates definite correlation between $|F_{r,t}|$ and $|R(\omega)|$ for each type of QPSs. Indeed, $|F_{r,t}|$ reach their peaks at frequencies ω_1 corresponding to the minima of the pump wave reflectivity, thus confirming that the pump wave refraction into stacks significantly influences the frequency mixing efficiency. It is also evident here that the CFG efficiency is hardly improved in the proximities of band edges for both QPSs and periodic stacks.

Comparison of $|F_{r,t}|$ for the stacks of the three types shows that the peak CFG efficiency is higher in the QPSs than in the periodic structure. For example, the $/F_r|$ peak at $\omega_1 = 2.38 \times 10^{13}$ s^{-1} in Thue-Morse QPS Q_5 (Figure 3b) is about 2.2 times higher than in the periodic stack (Figure 3c). Such enhanced CFG is attributed to more favourable conditions for the phase synchronism in the QPS and *local* field intensification in individual constituent layers due to the stack composition. Additionally, the internal refraction of pump waves and concurrent constructive interference of the generated combinatorial frequencies ω_3 in the entire stack lead to cumulative growth of the peak emission in the forward direction (note that almost all peaks of $|F_t|$ in Figure 3 are higher than $|F_r|$ peaks). Examination of the $|F_{r,t}|$ angular dependences has also proven that $|F_{r,t}|$ reach their maxima at the reflectance $|R(\omega)|$ minima of not only pump waves but also the waves of combinatorial frequency ω_3. This implies that the CFG efficiency can be further increased by optimising combinations of the incidence angles, layer parameters, and the stack composition.

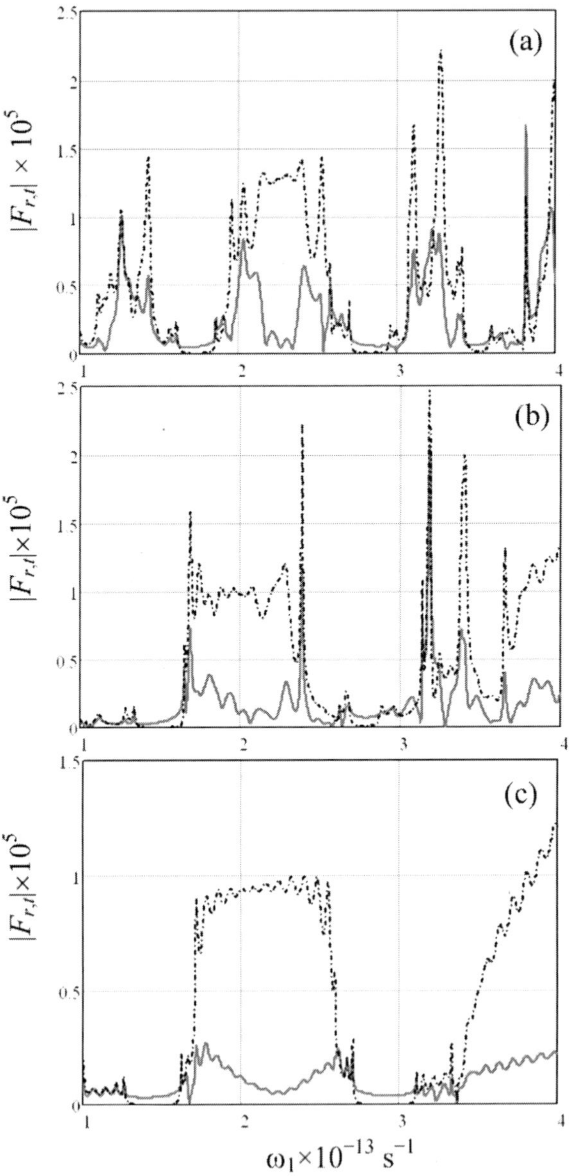

Figure 3. Intensity of field radiated from QPSs at frequency $\omega_3 = \omega_1 + \omega_2$ in the reverse ($|F_r|$-solid red lines) and forward ($|F_t|$-dash-dot

black lines) directions of the z-axis: (**a**) Fibonacci S_8 stack at $\omega_2 = 2.01 \times 10^{13}$ s^{-1}; (**b**) Thue-Morse Q_5 stack at $\omega_2 = 1.73 \times 10^{13}$s^{-1} and (**c**) periodic stack containing 16 unit cells at $\omega_2 = 1.76 \times 10^{13}$ s^{-1}. All the stacks are illuminated by pump waves incident at $\Theta_{i1} = 30°$ and $\Theta_{i2} = 45°$.

3.2. Effect of Stack Composition and Layer Anisotropy

Both the analytical estimations and numerical results shown in Figure 2 and Figure 3 have demonstrated strong effect of the stack internal arrangements on both the reflectance and CFG. To illustrate it further, Figure 4 displays $|F_{r,t}|$ simulated for Fibonacci S_7 stack, containing 21 layer with d_B = 19 μm, and Thue-Morse Q_4 stack, containing 16 layers with d_B = 26 μm, which have the same overall thicknesses as Fibonacci S_8 and Thue-Morse Q_5 QPSs in Figure 3 but thicker constituent layers. Comparison of Figure 3 and Figure 4 shows that intensities $|F_{r,t}|$ of the combinatorial frequency ω_3 emitted from the stacks have almost the same peak magnitudes in both cases but the peaks occur at different frequencies ω_1. Such spectral deviations can be attributed to pump wave redistribution in the stack constituent layers, caused by the internal reflection and refraction at the layer interfaces, and to the changes in the phase coherence between the pump waves and combinatorial frequency waves generated in individual constituent layers.

In order to assess the effect of the dielectric layers' anisotropy on CFG, Fibonacci and Thue-Morse QPSs of several different orders q have been re-simulated with isotropic constituent layers. The scalar relative permittivities $\varepsilon_{A,B}$ and nonlinear susceptibilities $\chi_{A,B}$ have been set equal to the tensor components specified above: $\varepsilon_{xxA} = \varepsilon_{zzA} = 5.382$, $\chi_{xxzA} = \chi_{zxxA} = \chi_{zzzA} = 2.1 \times 10^{-7}$ m/V, $\varepsilon_{xxB} = \varepsilon_{zzB} = 1.4$, $\chi_{xxzB} = \chi_{zxxB} = \chi_{zzzB} = 2.82 \times 10^{-8}$ m/V. Comparison of the simulation results for the isotropic and anisotropic layers has shown that the peak intensities of CFG are generally higher in isotropic cases, especially in QPSs with lower order q. The effect of the layer anisotropy in Fibonacci and Thue-Morse stacks appeared to be different. For example, in Fibonacci S_7 stack with isotropic layers $|F_{r,t}|$ peaks are nearly twice as high as in the same stacks with anisotropic layers, whereas in a similar Thue-Morse Q_4 configuration, magnitudes of $|F_{r,t}|$ peaks have about the same magnitudes. These observations suggest that a combination of the stack composition and the constituent layer anisotropy may noticeably influence CFG, especially in QPSs of low orders q. When q increases at fixed stack thickness, the $|F_{r,t}|$ peak magnitudes become closer in the stacks with isotropic and anisotropic layers. The latter trend can be attributed to thinning of the constituent layers that leads to the dominant effect of the spatial anisotropy enhanced by finer stratification of the stacks.

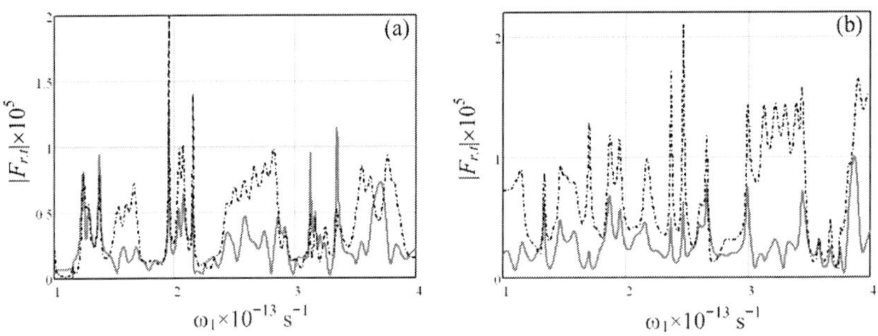

Figure 4. Intensity of field radiated from QPSs at frequency $\omega_3 = \omega_1 + \omega_2$ in the reverse ($|F_r|$-solid red lines) and forward ($|F_t|$-dash-dot black lines) directions of the z-axis: (**a**) Fibonacci S_7 stack at $\omega_2 = 2.02 \times 10^{13}$ s^{-1}; (**b**) Thue-Morse Q_4 stack at $\omega_2 = 1.76 \times 10^{13}$s^{-1}. All the stacks are illuminated by pump waves incident at $\Theta_{i1} = 30°$ and $\Theta_{i2} = 45°$.

3.3. Effect of Loss

To assess the effect of dissipation on CFG in QPSs, Fibonacci S_8 and Thue-Morse Q_5 stacks with the same layer parameters as in Figure 3 have been simulated taking into account the loss tangents of the constituent layers: tan $\delta_{xx,zz}$ = 0.01, 0.1. Comparison of the simulation results in Figure 3 and Figure 5 (lossless case) shows that the dielectric losses suppress sharp spikes of $|F_{r,t}|$ observed in the lossless QPSs and stronger affect emission in the forward direction than in the reverse direction of the z-axis [46].

The latter effect is particularly evident in Figure 5b,d where $|F_t|$ becomes nearly two orders of magnitude smaller than $|F_r|$. It is also important to note that not only magnitudes but also spectral content of the combinatorial frequencies emitted from the QPSs with imperfect layers qualitatively changes at higher losses.

These results demonstrate that CFG levels in the lossy QPSs are substantially decreased by dissipation of both pump waves and frequency mixing products. Their decays have different spatial scales, and dissimilar extinction rates of the pump waves may further reduce the CFG efficiency. Therefore, at relatively low losses (tan$\delta_{xx,zz}$ = 0.01), the spectral distributions of $|F_{r,t}|$ in Figure 5a,c still qualitatively resemble those for the corresponding lossless stacks in Figure 3. Conversely, at higher losses (tan$\delta_{xx,zz}$ = 0.1), $|F_{r,t}|$ distinctively differ because the frequency mixing products emitted from the stacks are predominantly generated in a few peripheral layers adjacent to the stack outer interfaces. Therefore $|F_r|$, primarily determined by CFG at the stack front interface where the pump waves are still weakly attenuated by losses, is more than two orders of magnitude higher than $|F_t|$ of the waves emitted from the other interface. Indeed, both the pump waves, travelling through the stack, and the mixing products, generated inside the stack, are strongly attenuated due to the layer losses. Therefore the $|F_t|$ magnitude is significantly lower than $|F_r|$ as illustrated by Figure 5b,d. In this case, the stack internal composition affects the CFG efficiency primarily through linear reflection, refraction, and pump wave extinction while the role of phase coherence becomes less prominent.

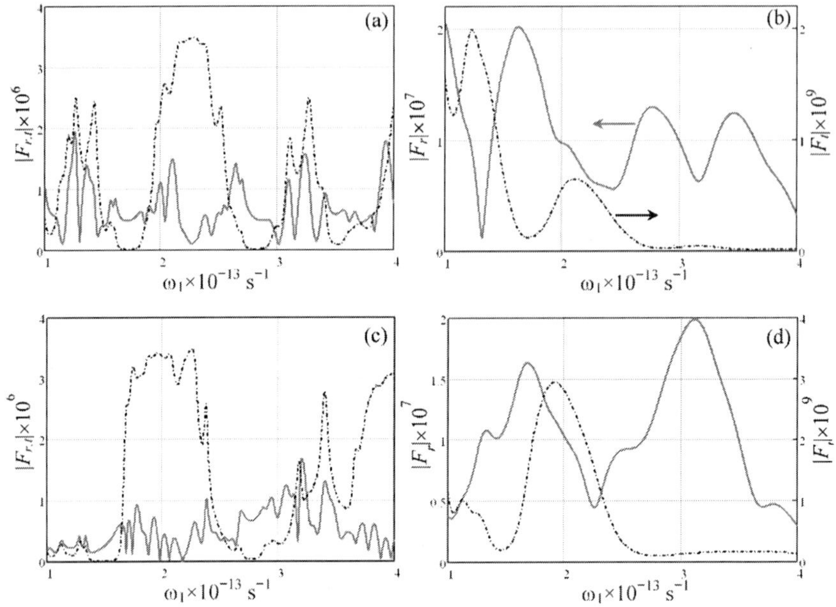

Figure 5. Intensity of field radiated at frequency $\omega_3 = \omega_1 + \omega_2$ from (**a**,**b**) Fibonacci $S_8$and (**c**,**d**) Thue-Morse Q_5 QPS in the reverse ($|F_r|$-solid red lines) and forward ($|F_t|$-dash-dot black lines) directions of the z-axis at $\omega_2 = 2.01 \times 10^{13}$ s^{-1}. The layer parameters are the same as in Figure 3a with added losses (**a**,**c**) - $\tan\delta_{xx,zz} = 0.01$ and (**b**,**d**) - $\tan\delta_{xx,zz} = 0.1$. The stack is illuminated by pump waves incident at $\Theta_{i1} = 30°$ and $\Theta_{i2} = 45°$.

4. CONCLUSIONS

Combinatorial frequency generation (CFG) in the three-wave mixing process has been studied in the quasi-periodic stacks of binary nonlinear dielectric layers arranged in accordance with Fibonacci and Thue-Morse sequences. The closed-form solutions of the nonlinear scattering problem for QPSs, illuminated by a pair of obliquely incident pump waves of frequencies ω_1 and ω_2, have been obtained in the approximation of weak nonlinearity using the harmonic balance method. The modified TMM has been adapted to analyse QPSs similarly to perturbed periodic stacks with defects at the specified positions. It has been shown that Fibonacci and Thue-Morse QPSs can be represented as cascades of "regular" and "defective" primitive cells. The types and locations of the "defective" cells have been determined and described in analytical form. The developed theory is illustrated by the examples of numerical simulations, and the features of CFG by Fibonacci and Thue-Morse QPSs are discussed.

The simulation results have demonstrated that the nonlinear scattering coefficients $|F_{r,t}(\omega_3)|$ of combinatorial frequency $\omega_3 = \omega_1 + \omega_2$ are strongly correlated with the stack linear reflectance $|R(\omega)|$. The latter effect is attributed to the fact that $|R(\omega)|$ determines magnitudes of both the pump waves of frequencies $\omega_{1,2}$, refracted into the stacks and engaged in the frequency mixing process, and the waves of combinatorial frequency ω_3 emitted from the stack. It has been found that the overall peak efficiency of CFG in QPSs is higher than in similar periodic stacks. The enhanced CFG in QPS is facilitated by the stack composition, which provides higher density of states, and improves the phase synchronism and local field intensification of the pump waves in the constituent layers. Comparison of CFG in Fibonacci and Thue-Morse stacks has shown that $|F_{r,t}(\omega_3)|$ peak magnitudes reach about the same levels in both types of QPSs but the peaks occur at different frequencies. At the same time, it has been observed that the layer anisotropy may affect the CFG efficiency in the low order QPSs, but its effect becomes marginal at the higher order QPSs. An extensive analysis of the CFG efficiency in lossy QPSs has shown that dissipation may qualitatively alter the frequency mixing process and change the $|F_r(\omega_3)|/|F_t(\omega_3)|$ ratio for several orders of magnitude. The latter effect is attributed to changes in the extinction scales and coherence lengths of pump waves and the CFG products.

ACKNOWLEDGMENTS

This work has been performed in the framework of the project PEARL supported by the FP7 Marie Curie IIF Grant 255110 and short visit grant from the European Science Foundation research networking program NEWFOCUS.

AUTHOR CONTRIBUTIONS

Oksana Shramkova and Alexander Schuchinsky jointly developed the theory, analysed results and wrote the paper. Oksana Shramkova has carried out the numerical simulations.

APPENDIX 1

To evaluate the fields of combinatorial frequency ω_3, it is necessary first to determine amplitudes coefficients $D^{n\pm}_{1j,2j}$ in Equation (4) for the waves generated inside each constituent layer of QPS. In the non-depleting wave approximation, this is accomplished by the harmonic balance method, which allows $D^{n\pm}_{1j,2j}(\omega_1, \omega_2)$ to be explicitly related to amplitudes $\mathcal{B}^{n\pm}_j(\omega_{1,2})$ of the pump waves of frequencies ω_1 and ω_2 refracted into each layer [40]. In contrast to regular periodic stacks, evaluation of $\mathcal{B}^{n\pm}_j(\omega)$ in QPSs is not straightforward and usually requires direct multiplication of the transfer matrices of all layers one by one. An alternative approach, based upon the QPS decomposition in "regular" and "defective" unit cells, is outlined in Section 2 and further elaborated here.

In the case of Fibonacci QPS defined in Section 2, the periodicity of the regular cells $\{AB\}$ is perturbed by additional layers of type A, which form the "defective" cells $\{AAB\}$ with layer A in front of the regular cells. The "regular" and "defective" cells are the primitive cells of any Fibonacci QPS S_q of order q. For example, in Fibonacci QPS S_5 = {AB AAB AB A}, the 2nd cell is defective, *i.e.*, it contains a layer A doublet of thickness $d'_A = 2d_A$. The regular and defective cells have the same layer sequence but different thicknesses: $d = d_A + d_B$ and $d' = d'_A + d_B$, respectively. Additionally, at odd q an extra layer A appears at the stack end, cf. S_5 above and S_6 in Section 2.

The overall number N_q of primitive cells (regular and defective) in Fibonacci QPS of order q is:

$$N_q = \begin{cases} \dfrac{\Phi_{q+1} - \Gamma_q}{2}, & q-\text{ even} \\ \dfrac{\Phi_{q+1} - \Gamma_q - 1}{2}, & q-\text{ odd} \end{cases} \tag{A1}$$

where $\Phi_q = \Phi_{q-1} + \Phi_{q-2}$ is Fibonacci number and $\Phi_1 = \Phi_2 = 1$; Γ_q is the number of defective cells: $\Gamma_q = 0$ at $q \leq 3$, and at $q \geq 4$.

$$\Gamma_q = \begin{cases} \dfrac{\left(\dfrac{3+\sqrt{5}}{2}\right)^{\frac{q}{2}-1} - \left(\dfrac{3-\sqrt{5}}{2}\right)^{\frac{q}{2}-1}}{\sqrt{5}}, & q-\text{ even} \\ \Gamma_{q+1} - \Gamma_{q-1} - 1, & q-\text{ odd} \end{cases} \tag{A2}$$

The positions of additional A layers in Fibonacci QPS of order q, can be determined with the aid of a row-matrix $\hat{P}_q$ of length Φ_{q+1}, composed of 0's and 1's. The 1's are located only in the columns corresponding to the first A layer of the doublets. At $q \geq 4$, $\hat{P}_q$ is defined by the recurrence relations:

$$\hat{P}_q = \hat{P}_{q-1} + \hat{P}_{q-2}\hat{\Psi}\left(\Phi_q\right) + \begin{cases} \hat{u}\left(\Phi_q\right), & q-\text{even} \\ 0, & q-\text{odd} \end{cases} \tag{A3}$$

where $\hat{u}\left(\Phi_q\right) = \left\{\delta_{i,\Phi_q}\right\}$ is a row-matrix with 1's in the Φ_q column only, $\delta_{i,i'}$ is Kronecker delta; $\hat{\Psi}\left(\Phi_q\right) = \left\{\delta_{i+\Phi_q,j}\right\}$ is a square Toeplitz matrix with 1's only at the secondary diagonal offset for Φ_q from the main diagonal. Then the serial number $v_i(q)$, i = 1,2, ... Γ_q of each defective cell in the stack can be deduced from $\hat{P}_q$ evaluated recursively in Equation (A3) for Fibonacci QPS of arbitrary order q:

$$\nu_i(q) = \frac{o_i(q) - i}{2} + 1, \tag{A4}$$

where $o_i(q)$ is the column number of the ith 1's in $\hat{P}_q$.

Thue-Morse QPS can be treated in a similar manner by representing a stack as a cascade of equal number of the regular $\{AB\}$ and defective $\{BA\}$ cells of the same thickness. The total number of primitive cells here is $N_q = 2^{q-1}$. The positions of the regular and defective cells, determined by their serial number n in the stack, can be deduced using the following recurrence relation at $n \geq 3$:

$$t_n = \begin{cases} 1 - t_{n-1}, & n - \text{even} \\ t_{\left(\frac{n-1}{2}+1\right)}, & n - \text{odd} \end{cases} \tag{A5}$$

where $t_n = 0$ represents the regular cells $\{AB\}$ and $t_n = 1$ – defective cells $\{BA\}$; $t_1 = 0$, $t_2 = 1$.

Once the positions of the regular and defective cells in QPS are determined, TMM can be applied to evaluating the pump wave scattering characteristics and fields inside each layer. Magnetic field $H_y(\omega_p, x, z)$ of TM wave of frequency ω_p outside the stack has the following form:

$$H_y(\omega_p, x, z) = \mathcal{A}_p e^{-i\omega_p t + ik_{xp}x} \begin{cases} e^{ik_{za}^{(p)}z} + R(\omega_p) e^{-ik_{za}^{(p)}z}, & z \leq 0 \\ T(\omega_p) e^{ik_{za}^{(p)}z}, & z \geq L_q \end{cases} \tag{A6}$$

and inside a layer of type $j = A, B$ in an nth primitive cell

$$H_{yj}^{(n)}(\omega_p, x, z) = \left(\mathcal{B}_j^{n+}(\omega_p) e^{ik_{zj}^{(p)}z} + \mathcal{B}_j^{n-}(\omega_p) e^{-ik_{zj}^{(p)}z} \right) e^{-i\omega_p t + ik_{xp}x}, \tag{A7}$$

where $k_{xp}=k_p\sqrt{\varepsilon_a}\sin\Theta_{ip}$, $k_{za}^{(p)}=k_p\sqrt{\varepsilon_a}\cos\Theta_{ip}$ and $k_{zj}^{(p)}=\sqrt{\left(k_p^2-\frac{k_{xp}^2}{\varepsilon_{zzj}}\right)\varepsilon_{xxj}}$ are the transverse and longitudinal wavenumbers outside the stack and inside the layers at frequencies ω_p, p = 1,2,3; $k_p = \omega_p/c$, c is the speed of light; A_p are amplitudes of the incident waves. $B_j^{n\pm}(\omega_p)$ are amplitude coefficients of the field in individual layers.

The stack reflection $R(\omega_p)$ and transmission $T(\omega_p)$ coefficients are obtained by interrelating the fields at the stack outer interfaces with the aid of the transfer matrix $\hat{M}_q(\omega_p)$:

$$\begin{pmatrix} H_{yA}(\omega_p,x,0) \\ E_{xA}(\omega_p,x,0) \end{pmatrix} = \hat{M}_q(\omega_p)\begin{pmatrix} H_{ys}(\omega_p,x,L_q) \\ E_{xs}(\omega_p,x,L_q) \end{pmatrix} \tag{A8}$$

where subscript s in RHS identifies the type of the last layer (A or B), which depends on the QPS type and order q:

$$\text{Fibonacci sequence: } s=\begin{cases} A, & q-\text{odd} \\ B, & q-\text{even} \end{cases}$$

$$\text{Thue-Morse sequence: } s=\begin{cases} A, & q-\text{even} \\ B, & q-\text{odd} \end{cases}$$

Making use of Equations (A6) and (A8) to satisfy the continuity conditions for the tangential fields at the stack outer interfaces, we obtain the QPS reflection and transmission coefficients:

$$\begin{aligned} R(\omega_p) &= \frac{M_{q11}(\omega_p)+\frac{k_{za}^{(p)}}{\varepsilon_a k_p}M_{q12}(\omega_p)-\frac{\varepsilon_a k_p}{k_{za}^{(p)}}M_{q21}(\omega_p)-M_{q22}(\omega_p)}{M_{q11}(\omega_p)+\frac{k_{za}^{(p)}}{\varepsilon_a k_p}M_{q12}(\omega_p)+\frac{\varepsilon_a k_p}{k_{za}^{(p)}}M_{q21}(\omega_p)+M_{q22}(\omega_p)} \\ T(\omega_p) &= \frac{2}{M_{q11}(\omega_p)+\frac{k_{za}^{(p)}}{\varepsilon_a k_p}M_{q12}(\omega_p)+\frac{\varepsilon_a k_p}{k_{za}^{(p)}}M_{q21}(\omega_p)+M_{q22}(\omega_p)} \end{aligned} \tag{A9}$$

The transfer matrices $\hat{M}_q(\omega_p)$ are evaluated by TMM at each frequency ω_p. For Fibonacci QPS, $\hat{M}_q(\omega_p)$ is defined recursively at $q \geq 2$.

$$\hat{M}_q(\omega_p) = \hat{M}_{q-1}(\omega_p)\hat{M}_{q-2}(\omega_p)$$

where $\hat{M}_0(\omega_p) = \bar{m}_{LB}(d_B, \omega_p)$, $\hat{M}_1(\omega_p) = \bar{m}_{LA}(d_A, \omega_p)$; $\bar{m}_{LA}(d_A, \omega_p)$ and $\bar{m}_{LB}(d_B, \omega_p)$ are the transfer matrices of layers $A$and B, respectively.

For Thue-Morse QPS at $q \geq 1$:

$$\hat{M}_q(\omega_p) = \hat{M}_{q-1}(\omega_p)\hat{M}'_{q-1}(\omega_p)$$
$$\hat{M}'_q(\omega_p) = \hat{M}'_{q-1}(\omega_p)\hat{M}_{q-1}(\omega_p)$$

with $\hat{M}_0(\omega_p) = \bar{m}_{LA}(d_A, \omega_p)$, $\hat{M}'_0(\omega_p) = \bar{m}_{LB}(d_B, \omega_p)$.

Once $R(\omega_p)$ and $T(\omega_p)$ are determined, amplitude coefficients $\mathcal{B}_j^{n\pm}(\omega_p)$ can be progressively deduced for each layer using the transfer matrices. In contrast to regular periodic structures, the direct evaluation of $\mathcal{B}_j^{n\pm}(\omega_p)$ for QPS is rather involved. The approach developed in this paper, where the QPSs are formed by cascading the regular and defective primitive cells, dramatically simplifies the analysis. The conventional procedure utilised for the periodic stacks can be adopted here as long as the positions of the defective cells are determined. Then the field amplitudes $\mathcal{B}_j^{n\pm}(\omega_p)$ in each layer can be expressed in the following form

$$\begin{aligned}\mathcal{B}_j^{n\pm}(\omega_p) = &\left(\eta_{j11}^{(n)}(\omega_p) \pm \frac{k_p}{k_{zL,j}^{(p)}}\varepsilon_{xy}\eta_{j21}^{(n)}(\omega_p)\right)\left(1+R(\omega_p)\right) \\ &+\frac{k_{za}^{(p)}}{k_p\varepsilon_a}\left(\eta_{j12}^{(n)}(\omega_p) \pm \frac{k_p}{k_{zL,j}^{(p)}}\varepsilon_{xy}\eta_{j22}^{(n)}(\omega_p)\right)\left(1-R(\omega_p)\right)\end{aligned} \quad \text{(A10)}$$

where $\bar{\eta}_j^{(n)}$ are the transfer matrices of a subset of $(n - 1)$ primitive cells preceding the layer of type $j = A, B$ in the nth primitive cell:

$$\begin{aligned} \bar{\eta}_A^{(n)}(\omega_p) &= \prod_{\ell=1}^{n-1} \bar{\kappa}^{(\ell)}(\omega_p), \\ \bar{\eta}_B^{(n)}(\omega_p) &= \bar{\eta}_A^{(n)}(\omega_p)\bar{\kappa}_L^{(n)}(\omega_p) \end{aligned} \tag{A11}$$

where

$$\bar{\kappa}^{(\ell)}(\omega_p) = \begin{cases} \bar{m}_{LA}(d_A,\omega_p)\bar{m}_{LB}(d_B,\omega_p) - \text{regular cells in Fibonacci QPS: } \ell \neq \nu_i(q) \text{ and} \\ \qquad \text{Thue-Morse QPS: } t_\ell = 0 \\ \bar{m}_{LA}(d'_A,\omega_p)\bar{m}_{LB}(d_B,\omega_p) - \text{defective cells in Fibonacci QPS: } \ell = \nu_i(q);\ i = 1,2,\ldots,\Gamma_q \\ \bar{m}_{LB}(d_B,\omega_p)\bar{m}_{LA}(d_A,\omega_p) - \text{defective cells in Thue-Morse QPS: } t_\ell = 1 \end{cases}$$

and

$$\bar{\kappa}_L^{(n)}(\omega_p) = \begin{cases} \hat{m}_{LA}(d_A,\omega_p) - \text{regular cells in Fibonacci QPS: } n \neq \nu_i(q) \text{ and} \\ \qquad \text{Thue-Morse QPS: } t_n = 0 \\ \bar{m}_{LA}(d'_A,\omega_p) - \text{defective cells in Fibonacci QPS: } n = \nu_i(q);\ i = 1,2,\ldots,\Gamma_q \\ \bar{m}_{LB}(d_B,\omega_p) - \text{defective cells in Thue-Morse QPS: } t_n = 1 \end{cases}$$

APPENDIX 2

The TM wave magnetic field $H_y(\omega_3)$ of frequency ω_3 emitted from the stack at combinatorial frequency ω_3 has the form:

$$H_y^a(\omega_3) = e^{-i\omega_3 t + ik_{x3}x} \begin{cases} F_r e^{-ik_{za}^{(3)}z}, & z \le 0 \\ F_t e^{ik_{za}^{(3)}z}, & z \ge L_q \end{cases}, \tag{A12}$$

where $k_{za}^{(3)} = \sqrt{k_3^2\varepsilon_a - k_{x3}^2}$ is the longitudinal wave number in the surrounding homogeneous media. To evaluate the field amplitudes

$F_{r,t}$ in Equation (A12), it is necessary to relate the fields of the nonlinear products generated inside the stack to the field emitted from the QPS. This is accomplished by enforcing the tangential field continuity at the stack external interfaces and applying the modified TMM [40] to find the fields inside the stack. The fields at the stack outer interfaces of Fibonacci QPS are related as follows

$$\begin{pmatrix} H_{yA}^{(1)}(\omega_3,x,0) \\ E_{xA}^{(1)}(\omega_3,x,0) \end{pmatrix} = \hat{M}_q(\omega_3) \begin{pmatrix} H_{ys}^{(N_q)}(\omega_3,x,L_q) \\ E_{xs}^{(N_q)}(\omega_3,x,L_q) \end{pmatrix} + \sum_{n=1}^{N_q} \left[\hat{\eta}_B^{(n)}(\omega_3) \begin{pmatrix} \tau_A^{(n)} \\ \xi_A^{(n)} \end{pmatrix} + \hat{\eta}_A^{(n+1)}(\omega_3) \begin{pmatrix} \tau_B^{(n)} \\ \xi_B^{(n)} \end{pmatrix} \right] + \begin{cases} \hat{\eta}_{N_q}(\omega_3)\hat{m}_{LA}(d_A,\omega_3) \begin{pmatrix} \tau_A^{(N_q+1)} \\ \xi_A^{(N_q+1)} \end{pmatrix}, & q\text{ - odd}, \\ 0, & q\text{ - even}, \end{cases} \tag{A13}$$

where N_q is the number of primitive cells in the stack; $\hat{\eta}_{N_q}(\omega_3) = \hat{M}_q(\omega_3)$ and $\hat{\eta}_j^{(n)}$ (ω_3) are defined in (A1.11); $\tau_j^{(n)}$ and $\xi_j^{(n)}$ at $j = A$, B are:

$$\begin{aligned} \tau_j^{(n)} &= D_{1j}^{n+}\sigma_{1j}^{n+}(k_{zj}^{+}) + D_{2j}^{n+}\sigma_{1j}^{n-}(k_{zj}^{+}) + D_{1j}^{n-}\sigma_{1j}^{n+}(k_{zj}^{-}) + D_{2j}^{n-}\sigma_{1j}^{n-}(k_{zj}^{-}) \\ \xi_j^{(n)} &= D_{1j}^{n+}\sigma_{2j}^{n+}(k_{zj}^{+}) + D_{2j}^{n+}\sigma_{2j}^{n-}(k_{zj}^{+}) + D_{1j}^{n-}\sigma_{2j}^{n+}(k_{zj}^{-}) + D_{2j}^{n-}\sigma_{2j}^{n-}(k_{zj}^{-}), \end{aligned} \tag{A14}$$

$D_{1j,2j}^{n\pm}$ introduced in Equation (4) are proportional to the pump wave amplitude products: $\mathcal{B}_j^{n\pm}(\omega_1)\mathcal{B}_j^{n\pm}(\omega_2)$ [40].

$$\sigma_{1j}^{n\pm}(\upsilon) = \left(\cos k_{zj}^{(3)} d_j^{(n)} - e^{\pm i\upsilon d_j^{(n)}}\right) \pm i\frac{\upsilon}{k_{zj}^{(3)}} \sin k_{zj}^{(3)} d_j^{(n)};$$

$$\sigma_{2j}^{n\pm}(\upsilon) = \frac{\upsilon}{k_3\varepsilon_{xxj}} \left[i\frac{k_{zj}^{(3)}}{\upsilon} \sin k_{zj}^{(3)} d_j^{(n)} \pm \left(\cos k_{zj}^{(3)} d_j^{(n)} - e^{\pm i\upsilon d_j^{(n)}}\right) \right],$$

$$d_B^{(n)} = d_B, \qquad d_A^{(n)} = \begin{cases} d_A - \text{regular cells} \\ d'_A - \text{defective cells} \end{cases}.$$

Similarly, the fields at the stack outer interfaces of Thue-Morse QPS are related as follows:

$$\begin{pmatrix} H_{yA}^{(1)}(\omega_3,x,0) \\ E_{xA}^{(1)}(\omega_3,x,0) \end{pmatrix} = \bar{M}_q(\omega_3) \begin{pmatrix} H_{ys}^{(N_q)}(\omega_3,x,L_q) \\ E_{xs}^{(N_q)}(\omega_3,x,L_q) \end{pmatrix} + \sum_{n=1}^{N_q} \left[\bar{\eta}_B^{(n)}(\omega_3) \begin{pmatrix} \tau_1^{(n)} \\ \xi_1^{(n)} \end{pmatrix} + \bar{\eta}_A^{(n+1)}(\omega_3) \begin{pmatrix} \tau_2^{(n)} \\ \xi_2^{(n)} \end{pmatrix} \right]$$

(A15)

where $\bar{\eta}_j^{(n)}(\omega_3)$ are defined in Equation (A11); $\tau_{1,2}^{(n)}$ and $\xi_{1,2}^{(n)}$ are as follows

$$\tau_1^{(n)} = \begin{cases} \tau_A^{(n)} - \text{regular cells} \\ \tau_B^{(n)} - \text{defective cells} \end{cases} \qquad \tau_2^{(n)} = \begin{cases} \tau_B^{(n)} - \text{regular cells} \\ \tau_A^{(n)} - \text{defective cells} \end{cases}$$

$$\xi_1^{(n)} = \begin{cases} \xi_B^{(n)} - \text{regular cells} \\ \xi_A^{(n)} - \text{defective cells} \end{cases} \qquad \xi_2^{(n)} = \begin{cases} \xi_A^{(n)} - \text{regular cells} \\ \xi_B^{(n)} - \text{defective cells} \end{cases}$$

$\tau_j^{(n)}$ and $\xi_j^{(n)}$, $j = A, B$ have the form of Equation (A14) except the layer *A* thickness remaining the same in both regular and defective cells, *i.e.*, $d_A^{(n)} = d_A$.

Combining Equation (A12) with Equations (A14) or (A15) in the boundary conditions for the field continuity at the stack external interfaces gives amplitudes $F_{r,t}$ of the waves emitted from the QPS at combinatorial frequency ω_3:

$$F_r = \left(\frac{k_3 \varepsilon_a}{k_{za}^{(3)}} \bar{\eta}_{N_q}(\omega_3)_{2,1} + \bar{\eta}_{N_q}(\omega_3)_{2,2} \right) \lambda_1 - \left(\frac{k_3 \varepsilon_a}{k_{za}^{(3)}} \bar{\eta}_{N_q}(\omega_3)_{1,1} + \bar{\eta}_{N_q}(\omega_3)_{1,2} \right) \lambda_2,$$

$$F_t = -\left(\lambda_1 + \lambda_2 \frac{k_3 \varepsilon_a}{k_{za}^{(3)}} \right),$$

(A16)

where $\bar{\eta}_j^{(n)}(\omega_3)$ are defined in (A.11) and $\lambda_{1,2}$ are

- for Fibonacci QPS:

$$\lambda_r = \frac{1}{\Delta}\sum_{n=2}^{N_q}\left[\hat{\eta}_B^{(n)}(\omega_3)_{r,1}\,\tau_A^{(n)} + \hat{\eta}_B^{(n)}(\omega_3)_{r,2}\,\xi_A^{(n)} + \hat{\eta}_A^{(n+1)}(\omega_3)_{r,1}\,\tau_B^{(n)} + \hat{\eta}_A^{(n+1)}(\omega_3)_{r,2}\,\xi_B^{(n)}\right] + \begin{cases}\left[\hat{\eta}_{N_q}\hat{m}_{LA}(d_A,\omega_3)\right]_{r,1}\tau_A^{(N_q+1)} + \left[\hat{\eta}_{N_q}\hat{m}_{LA}(d_A,\omega_3)\right]_{r,2}\xi_A^{(N_q+1)}, & q\text{ - odd}\\ 0, & q\text{ - even}\end{cases} \quad r = 1,2;$$

- for Thue-Morse QPS:

$$\lambda_r = \frac{1}{\Delta}\sum_{n=2}^{N_q}\left[\hat{\eta}_B^{(n)}(\omega_3)_{r,1}\,\tau_1^{(n)} + \hat{\eta}_B^{(n)}(\omega_3)_{r,2}\,\xi_1^{(n)} + \hat{\eta}_A^{(n+1)}(\omega_3)_{r,1}\,\tau_2^{(n)} + \hat{\eta}_A^{(n+1)}(\omega_3)_{r,2}\,\xi_2^{(n)}\right], \quad r = 1,2$$

$$\Delta = \hat{\eta}_{N_q}(\omega_3)_{1,1} + \hat{\eta}_{N_q}(\omega_3)_{2,2} + \frac{k_{za}^{(3)}}{k_3\varepsilon_a}\hat{\eta}_{N_q}(\omega_3)_{1,2} + \frac{k_3\varepsilon_a}{k_{za}^{(3)}}\hat{\eta}_{N_q}(\omega_3)_{2,1}$$

CONFLICTS OF INTEREST

The authors declare no conflict of interest.

REFERENCES AND NOTES

1. Valy Vardeny, Z.; Nahata, A.; Agrawal, A. Optics of photonic quasicrystals. *Nat. Photon.* **2013**, *7*, 177–187.

2. Merlin, R.; Bajema, K.; Clarke, R.; Juang, F.-Y.; Bhattacharya, P.K. Quasiperiodic GaAs-AlAs Heterostructures. *Phys. Rev. Lett.* **1985**, *55*, 1768–1770.

3. Pelster, R.; Gasparian, V.; Nimtz, G. Propagation of plane waves and of waveguide modes in quasiperiodic dielectric heterostructures. *Phys. Rev. E* **1997**, *55*, 7645–7655.

4. Liu, N. Propagation of light waves in Thue-Morse dielectric multilayers. *Phys. Rev. B* **1997**, *55*, 3543–3547.

5. Maciá, E. Hierarchical description of phonon dynamics on finite Fibonacci superlattices. *Phys. Rev. B* **2006**, *73*.

6. Albuquerque, E.L.; Cottam, M.G. *Polaritons in Periodic and Quasiperiodic Structures*; Elsevier: Amsterdam, The Netherland, 2004.

7. Zoorob, M.E.; Chariton, M.D.B.; Parker, G.J.; Baumberg, J.J.; Netti, M.C. Complete photonic bandgaps in 12-fold symmetric quasicrystals. *Nature* **2000**, *404*, 740–743.

8. Cojocaru, E. Omnidirectional reflection from finite periodic and Fibonacci quasi-periodic multilayers of alternating isotropic and birefringent thin films. *Appl. Opt.* **2002**, *41*, 747–755.

9. Barriuso, A.; Monzón, J.; Sánchez-Soto, L.; Felipe, A. Comparing omnidirectional reflection from periodic and quasiperiodic one-dimensional photonic crystals. *Opt. Exp.* **2005**, *13*, 3913–3920.

10. Freedman, B.; Bartal, G.; Segev, M.; Lifshitz, R.; Christoet, D.N. Wave and defect dynamics in nonlinear photonic quasicrystals. *Nature* **2006**, *440*, 1166–1169.

11. El Hassouani, Y.; Aynaou, H.; El Boudouti, E.H.; Djafari-Rouhani, B.; Akjouj, A.; Velasco, V.R. Surface electromagnetic waves in Fibonacci superlattices: Theoretical and experimental results. *Phys. Rev. B* **2006**, *74*.

12. Moretti, L.; Rea, I.; Rotiroti, L.; Rendina, I.; Abbate, G.; Marino, A.; de Stefano, L. Photonic band gaps analysis of Thue-Morse multilayers made of porous silicon. *Opt. Exp.* **2006**, *14*, 6264–6272.

13. Makarava, L.N.; Nazarov, M.M.; Ozheredov, I.A.; Shkurinov, A.P.; Smirnov, A.G.; Zhukovsky, S.V. Fibonacci-like photonic structure for femtosecond pulse compression. *Phys. Rev. E* **2007**, *75*.

14. Tuz, V. Optical properties of a quasi-periodic generalized Fibonacci structure of chiral and material layers. *JOSA B* **2009**, *26*, 627–632.

15. Reyes-Gómez, E.; Raigoza, N.; Cavalcanti, S.B.; de Carvalho, C.A.A.; Oliveira, L.E. Plasmon polaritons in photonic metamaterial Fibonacci superlattices. *Phys. Rev. B* **2010**, *81*.

16. Pang, X.-N.; Dong, J.-W.; Wang, H.-Z. Photonic localization of interface modes at the boundary between metal and Fibonacci quasiperiodic structure. *JOSA B* **2010**, *27*, 2009–2013.

17. Olkhovskiy, Y.A.; Shramkova, O.V. Electromagnetic wave transmission and reflection by a quasiperiodic layered semiconductor structure. *Phys. B Condens. Matter* **2011**, *406*, 1415–1419.

18. Dolev, I.; Volodarsky, M.; Porat, G.; Arie, A. Multiple coupling of surface plasmons in quasiperiodic gratings. *Opt. Lett.* **2011**, *36*, 1584–1586.

19. Hsueh, W.J.; Wun, S.J.; Lin, Z.J.; Cheng, Y.H. Features of the perfect transmission in Thue-Morse dielectric multilayers. *J. Opt. Soc. Am. B* **2011**, *28*, 2584–2591.

20. Hsueh, W.J.; Chang, C.H.; Cheng, Y.H.; Wun, S.J. Effective Bragg conditions in a one-dimensional quasicrystal. *Opt. Expr.* **2012**, *20*, 26618–26623.

21. Kruk, S.S.; Helgert, C.; Decker, M.; Staude, I.; Menzel, C.; Etrich, C.; Rockstuhl, C.; Jagadish, C.; Pertsch, T.; Neshev, D.N.; *et al.* Optical metamaterials with quasicrystalline symmetry: Symmetry-induced optical isotropy. *Phys. Rev. B* **2013**, *88*.

22. Savoia, S.; Castaldi, G.; Galdi, V. Optical nonlocality in multilayered hyperbolic metamaterials based on Thue-Morse superlattices. *Phys. Rev. B* **2013**, *87*.

23. Xu, S.; Zhu, Y.; Wang, L.; Yang, P.; Chu, P.K. Passband and defective bands in photonic and quasi-crystals. *J. Opt. Soc. Am. B* **2014**, *31*, 664–671.

24. Gong, Y.; Liu, X.; Wang, L.; Lu, H.; Wang, G. Multiple responses of TPP-assisted near-perfect absorption in metal/Fibonacci quasiperiodic photonic crystal. *Opt. Expr.* **2011**, *19*, 9759–9769.

25. Feng, J.; Zhu, Y.; Ming, N. Harmonic generations in an optical Fibonacci superlattice. *Phys. Rev. B* **1990**, *41*, 5578–5582.

26. Zhu, S.; Zhu, Y.; Ming, N. Quasi-phase-matched third-harmonic generation in a quasi-periodic optical superlattice.*Science* **1997**, *278*, 843–846.

27. Fradkin-Kashi, K.; Arie, A. Multiple-wavelength quasi-phase-matched nonlinear interactions. *IEEE J. Quant. Electron.***1999**, *35*, 1649–1656.

28. Chern, G.-W.; Chang, J.-F.; Wang, L.A. Modeling of nonlinear pulse propagation in periodic and quasi-periodic binary long-period fiber gratings. *JOSA B* **2002**, *19*, 1497–1508.

29. Huang, T.; Huang, D. Interplay between optical nonlinearity and localization in a finite disordered Fibonacci chain.*Phys. Rev. B* **2007**, *76*.

30. Lifshitz, R.; Arie, A.; Bahabad, A. Photonic quasicrystals for nonlinear optical frequency conversion. *Phys. Rev. Lett.***2005**, *95*.

31. Werchner, M.; Schafer, M.; Kira, M.; Koch, S.W.; Sweet, J.; Olitzky, J.D.; Hendrickson, J.; Richards, B.C.; Khitrova, G.; Gibbs, H.M.; *et al.* One dimensional resonant Fibonacci quasicrystals: Noncanonical linear and canonical nonlinear effects. *Opt. Expr.* **2009**, *17*, 6813–6828.

32. Vasconcelos, M.S.; Mauriz, P.W.; Albuquerque, E.L. Optical filters based in quasiperiodic photonic crystal.*Microelectron. J.* **2009**, *40*, 851–853.

33. Grigoriev, V.; Biancalana, F. Bistability, multistability and non-reciprocal light propagation in Thue-Morse multilayered structures. *New J. Phys.* **2010**, *12*.

34. Zhukovsky, S.V.; Smirnov, A.G. All-optical diode action in asymmetric nonlinear photonic multilayers with perfect transmission resonances. *Phys. Rev. A* **2011**, *83*.

35. Lotfi, E.; Jamshidi-Ghaleh, K.; Moslemi, F.; Masalehdan, H. Optical multistability in 1D photonic crystals with nonlinear Thue-Morse structure. *Appl. Phys. A* **2011**, *103*, 669–672.

36. Zhu, S.N.; Zhu, Y.Y.; Qin, Y.Q.; Wang, H.F.; Ge, C.Z.; Ming, N.B. Experimental realization of second harmonic generation in a Fibonacci optical superlattice of $LiTaO_3$. *Phys. Rev. Lett.* **1997**, *78*, 2752–2755.

37. Chen, Y.B.; Zhang, C.; Zhu, Y.Y.; Zhu, S.N.; Wang, H.T.; Ming, N.B. Optical harmonic generation in a quasi-phase-matched three-component Fibonacci superlattice $LiTaO_3$. *Appl. Phys. Lett.* **2001**, *78*, 577–579.

38. Sheng, Y.; Koynov, K.; Zhang, D. Collinear second harmonic generation of 20 wavelengths in a single two-dimensional decagonal nonlinear photonic quasi-crystal. *Opt. Commun.* **2009**, *282*, 3602–3606.

39. Blombergen, N. *Nonlinear Optics: A Lecture Note*; Benjamin: New York, NY, USA & Amsterdam, The Netherland, 1965.

40. Shramkova, O.V.; Schuchinsky, A.G. Nonlinear scattering by anisotropic dielectric periodic structures. *Adv. OptoElectron.* **2012**.

41. Jeong, Y.; Lee, B. Matrix analysis for layered quasi-phase-matched media considering multiple reflection and pump wave depletion. *IEEE J. Quant. Electron.* **1999**, *35*, 162–172.

42. Li, J.-J.; Li, Z.-Y.; Zhang, D.-Z. Second harmonic generation in one-dimensional nonlinear photonic crystals solved by the transfer matrix method. *Phys. Rev. E* **2007**, *75*.

43. Szczepański, P.; Osuch, T.; Jaroszewicz, Z. Modeling of amplification and light generation in one-dimensional photonic crystal using a multiwavelength transfer matrix approach. *Appl. Opt.* **2010**, *48*, 5401–5406.

44. Li, J.-J.; Li, Z.-Y.; Zhang, D.-Z. Nonlinear frequency conversion in two-dimensional nonlinear photonic crystals solved by a plane-wave-based transfer-matrix method. *Phys. Rev. B* **2008**, *77*.

45. Yariv, A.; Yeh, P. *Optical Waves in Crystals: Propagation and Control of Laser Radiation*; Wiley: New York, NY, USA, 1984.

46. The approximation of the non-depleting nonlinear process is used here. Therefore extinction of pump waves is caused by the dissipative losses only.

INDEX

L

M

N

O

P

Q

R

S

T

V

Y

Z